Heat Pump-Assisted Dryers

Heat Pump-Assisted Dryers: Optimum Sizing, Design, and Operation Rules identifies the optimum sizing, operation, and control strategies for hot air convective batch driers coupled with air-to-air mechanical vapor compression heat pumps. It evaluates the energy requirements, most efficient operation and control methods, drying performances, and environmental issues and considerations.

Covering the physical drying process of porous solid materials (fruits, vegetables, and wood), the book discusses heat conduction and moisture migration. It investigates the optimum coupling of thermodynamic systems, examining unproved drying techniques and combining solar energy and heat pump-assisted dryers. The book explores future R&D needs, including advanced drying technologies, low GWP refrigerants, and process integration.

This book is an essential resource for professionals working with heat pumps, industrial process engineers, manufacturers, and academic researchers studying advances in drying technologies and processes.

Heat Pump-Assisted Dryers

Optimum Sizing, Design, and Operation Rules

Vasile Minea

CRC Press
Taylor & Francis Group
Boca Raton London New York

CRC Press is an imprint of the
Taylor & Francis Group, an **informa** business

Designed cover image: MRS déshumidification

First edition published 2026
by CRC Press
2385 NW Executive Center Drive, Suite 320, Boca Raton FL 33431

and by CRC Press
4 Park Square, Milton Park, Abingdon, Oxon, OX14 4RN

CRC Press is an imprint of Taylor & Francis Group, LLC

ISBN: 978-1-041-02252-7 (hbk)
ISBN: 978-1-041-02254-1 (pbk)
ISBN: 978-1-003-61838-6 (ebk)

DOI: 10.1201/9781003618386

Typeset in Times
by KnowledgeWorks Global Ltd.

Contents

Preface

Drying of solid porous materials is one of the most common, but complex and energy-intensive processes existing in many industries. Its extensive study consists in theoretical/modeling/simulation works and real experimental measures.

The book *Heat Pump-Assisted Dryers: Optimum Sizing, Design, and Operation Rules* contains nine chapters intending at updating optimum coupling, sizing and design, as well as efficient operation and control rules to improve the energy efficiency and environmental sustainability for heat pump-assisted batch dryers.

It shows that over the last few decades, materials such as fruits, vegetables, and woods have been dried with heat pump-assisted dryers, but relatively few published R&D studies have provided complete and/or pertinent information about theoretical and experimental initial data, system sizing, running and control methods used, and/or reliable results. The book develops new pertinent concepts to avoid undesirable troubleshooting and/or failures of mechanical vapor compression heat pumps coupled with hot air convective batch drying enclosures, as well as poor and unreliable energy and drying performances.

The book also emphasizes that future R&D works should at least (i) use as simple as possible configurations coupling hot air convective batch dryers with air-to-air nonreversible mechanical vapor compression heat pumps—in other words, do not uselessly complicate the layouts of such coupled systems; (ii) indicate the durations of preheating steps and heat sources used; (iii) optimize the calculation of heat pumps' nominal/design dehumidification (evaporator thermal power) capacities in order to continuously match materials' dewatering rates with heat pumps' actual dehumidification capacities during the entire drying cycles; (iv) use efficient heat pump running strategies and optimum control rules; (v) specify, based on continuous measures, drying air dry-bulb and wet-bulb temperatures and relative and absolute humidity allowing setting the heat pumps' compressors running sequences; (vi) indicate the total quantities of moisture removed from dried materials; (vii) express modified specific moisture extraction ratios and modified specific energy consumptions based on the system (dryer plus heat pump) energy consumptions during the entire drying cycles.

The book warns professionals working with heat pump-assisted drying systems, such as industrial process engineers, manufacturers, and research and design personnel, about potential troubleshooting and wrong drying performances that can result from inadequate coupling, sizing, and/or control strategies of laboratory-scale and/or industrial-scale heat pump-assisted dryers.

The objective is to ensure heat pump-assisted dryers energy performances are similar to or higher than those of conventional drying methods, while generating energy-savings and enhancing the final quality of dried products such as fruits, vegetables, and woods.

Vasile Minea
Retired, Hydro-Québec Research Institute
Laboratory of Energy Technology, LTE (Québec), Canada

Author

Vasile Minea is a Ph.D. graduate of Bucharest Technical Construction University, Romania. He worked as university professor for over 15 years, teaching courses in hydraulic/fluid and HVAC (heating, ventilating, and air-conditioning) systems for civil, agricultural, and industrial processes and buildings, as well as thermodynamics, heat transfer, and refrigeration. From 1987, Dr. Minea worked for more than 30 years as a scientist researcher at the Hydro-Quebec Research Institute, Laboratory of Energy Technologies in Quebec, Canada. His research activity mainly focused on industrial refrigeration and heat pumps, heat recovery technologies in large-scale commercial/institutional HVAC systems, and various industrial processes (such as heat pump-assisted drying), low- and medium-temperature geothermal heat pumps, and medium-enthalpy power generation cycles. During this period, Dr. Minea collaborated with the Canadian and American heat pump industry and R&D drying community in developing laboratory- and industrial-scale experimental heat pump-assisted drying systems.

Acknowledgments

I gratefully acknowledge the publisher, CRC Press, for its sustained involvement, assistance, patience, and strong support throughout the entire process of writing this book about heat pump-assisted drying technology that contribute to lowering the world's global fossil energy consumption and greenhouse gas emissions.

Specifically, I would like to mention Allison Shatkin, Senior Editor, Books, Material Science & Chemical Engineering, and Sonia Tam, Editorial Assistant, both at Taylor & Francis Group, for their courteous, professional, and helpful assistance.

Vasile Minea

Retired, Hydro-Québec Research Institute

Laboratory of Energy Technology, LTE (Québec), Canada

1 Introduction

1.1 INTRODUCTION

Drying of hygroscopic and porous solid materials such as agro-foods (e.g., fruits and vegetables) and woods (e.g., softwoods and hardwood) is a complex, highly energy-intensive process involving coupled heat and moisture (mass) transfers (Chua et al. 2002; Mujumdar 2007). It occurs by supplying heat to wet materials and by vaporizing the liquid (water) contained inside. Worldwide, about 85% of industrial driers are of convective type with hot air as drying medium.

Owing to the rising energy prices and growing concerns about the global climate change, reducing energy consumption in drying industry has become a challenging task. Most of energy (heat) losses in industrial batch driers are due to the exhausted hot and moist air and, sometimes, due to poor thermal insulation of drying enclosures. Among the advanced heat recovery methods is the heat pump-assisted drying method that couples air-to-air mechanical vapor compression heat pumps (acting as cooling, dehumidifying, and reheating devices) with slightly modified conventional hot air convective batch drying chambers. Such a drying method recovers latent heat by condensing moisture from the hot and moist drying air. The recovered heat is recycled back to practically closed batch dryers at higher temperatures for reuse in the same drying cycles. Therefore, they have the potential to reduce a part of primary energy consumed for drying many valuable products. In addition, compared to traditional hot air convective batch drying, heat pump-assisted dryers can operate independent of the outside ambient air conditions. They are environment-friendly because water vapor and polluted and/or toxic gases are not exhausted into the atmosphere. Also, particularly in the wood drying industry, the water condensate can be recovered and disposed of, offering opportunities for chemical treatment and, sometimes, recovery of valuable substances. Despite such multiple technical, economical, and environmental advantages, the industrial applications of heat pump-assisted drying systems have not increased as expected in the past.

In the past few decades, several published works did not specify the quantities of dried materials, nominal sizes of mechanical vapor compression heat pumps, specific drying schedules (including preheating steps, material initial and final moisture contents, drying air wet-bulb temperatures, and even heat pump coefficients of performance or system drying performances) while reporting high energy performances. In addition, information about how the material dewatering vaporization rates were matched with heat pumps' actual moisture condensation rates was not indicated. Moreover, many published studies did not highlight the excess external heat rejected by heat pump-assisted drying systems, which seriously affected the accuracy of reported overall drying (energy) performances.

DOI: 10.1201/9781003618386-1

This book focuses on basic principles, such as dryer–heat pump coupling, and rules for optimum size, design, and operate and control of heat pump-assisted batch dryers. It suggests that a good understanding of both hot air convective batch dryers and air-to-air, nonreversible mechanical vapor compression heat pump technologies, followed by a strong collaboration between academia and industrial drying R&D communities, may ensure more successful future industrial implementations.

This book also notes that to achieve high drying performances and to be able to validate them, for a given heat pump-assisted dryer configuration, it is necessary to always provide information (among many others) on (i) material-specific drying kinetics; (ii) drier's main characteristics, such as the location (hot or cold weather, indoor, outdoor), volume (or mass) capacity, configuration of air circulation through both dryer and heat pump (fans, blowers, air vents), and percentage of dryer heat leakage rates (conduction, air ex- and infiltration); (iii) drying schedules (for each drying step, at least drying air dry- and wet-bulb temperatures, or air temperature depressions, and relative humidity); (iv) material input and output masses (or volumes), as well as initial and final moisture contents; (v) warming-up (preheating) requirements and type of supplementary (backup) heating (if applicable); and (vi) type of refrigerant used and maximum and minimum allowed discharge pressures and temperatures.

Chapter 2 presents the main laws (e.g., Fourier's law and Fick's first and second laws) involved in internal and external heat and moisture transfer mechanisms occurring during drying of porous solid materials such as fruits, vegetables, and woods. For steady-state, external, parallel moist air streams, the main thermal and hydraulic parameters are described that control the boundary conditions and affect the rate of moisture vaporization. Typical drying periods, such as warming-up, the constant and first and second falling rate steps are identified and briefly described.

Chapter 3 points out that drying products such as fruits, vegetables, and woods is a complex process involving simultaneous heat and moisture (mass) transfers. It is influenced by several internal (e.g., density, permeability, porosity, sorption–desorption characteristics, and thermophysical properties) and external (e.g., temperature, velocity, and relative humidity of drying medium) parameters, as well as typical drying schedules. The main conventional and advanced drying methods as well as several published experimental and theoretical results are also summarized.

Chapter 4 presents the green wood types, structures, and processing steps. It describes the main drying properties of wood (density, moisture content, fiber saturation point, and equilibrium moisture content), important aspects of drying at low-, medium- and high-temperatures, and the lumber board's internal and external heat transfer and moisture movement. It discusses the advantages of wood drying and summarizes the conventional drying methods (natural, solar, hot air convective, vacuum, and superheated vapor drying), main drying periods, and defects that may occur (such as shrinkage).

Chapter 5 critically reviews several published theoretical and experimental R&D studies on drying fruits, vegetables, and woods using hot air convective batch dryers coupled with electrically driven nonreversible air-to-air mechanical vapor compression heat pumps. It analyzes the relevance as well as the accuracy of proposed configurations and associated control methods. It also evaluates the pertinence and/

or validity of the reported (published) laboratory and in-field experimental results versus data provided, as well as the potential of industrial applications of several proposed systems.

Chapter 6 presents the important advantages and limitations of heat pump-assisted drying systems. It describes the basic thermodynamic cycle of single-stage, nonreversible, electric-driven subcritical air-to-air mechanical vapor compression heat pumps and the main characteristics of conventional hot air convective batch driers. It proposes to academia and industry drying and heat pump communities some optimum (efficient) coupling methods of hot air convective batch dryers with air-to-air mechanical vapor compression heat pumps, as well as several combined concepts using additional energy sources such as solar, radio frequency, and infrared. It identifies the main energy performance indicators of heat pump-assisted drying systems, and, in the case of heat pump-assisted wood drying systems, some of environmental issues.

After having efficiently coupled the two thermodynamic systems (i.e., hot air convective batch dryers and air-to-air mechanical vapor compression heat pumps), the next step involves optimum calculation of maximum, minimum, and average required dehumidification capacities of air-to-air mechanical heat pumps. For efficient operation and high drying performances, Chapter 7 points out that mechanical vapor compression heat pumps should run continually as long as possible during the drying cycles. To achieve this, adequate correlation between the materials' thermal dewatering rates and heat pumps' dehumidification capacities (i.e., evaporator thermal powers) during each step of drying cycles must be ensured. Besides, initial masses (or volumes) of dried materials, as well as the total quantities of moisture to remove, approximate (estimated) drying times, and the material required for final moisture contents should be carefully determined before the system design process.

Chapter 8 proposes (developed) some efficient operation and control rules of thermodynamic systems consisting of closed-loop refrigerant circuits coupled with drying air closed-loop circuits. Among these rules, the amount of evaporator refrigerant superheating helps to avoid refrigerant saturated vapor–liquid mixtures or even liquid drops to enter the compressor suction line. Another major rule requires appropriate setting of drying air wet-bulb temperatures to avoid too short on/off drying cycles. Chapter 8 also notes that efficient management of refrigerant migration throughout the heat pump piping and components could be provided by adequate quantity mass of refrigerant charge, by using refrigerant liquid valves and/or artificially cooled suction refrigerant accumulators. Finally, for medium- and high-temperature split-type heat pump-assisted dryers, an improved supplementary (and backup) heat supply technique is developed to avoid abnormal heat pump operation parameters.

Chapter 9 reviews the pertinence and reliability of existing and/or of new proposed heat pump-assisted drying concepts, as well as the correlation between the data provided and the results reported by several published R&D works. Some of the most oft-met design errors, operational drawbacks, missing data, and wrong experimental results are identified. Chapter 9 lists the most important requirements for future theoretical and experimental R&D works, as well as for journal and conference publications on heat pump-assisted systems. It warns professionals working with drying systems and heat pumps, industrial process engineers, manufacturers, and research and design personnel about potential problems that can result from inadequate coupling, sizing, and/or control strategies of laboratory as well as industrial-scale heat pump-assisted dryers.

REFERENCES

Chua, K.J., S.K. Chou, J.C. Ho, N.A. Hawlader. 2002. Heat pump drying: Recent developments and future trends. *Drying Technology* 20(8):1579–1610.

Mujumdar, A.S. 2007. *Handbook of Industrial Drying*, 3rd edition, Taylor & Francis Group, Oxfordshire, UK.

2 Drying of Porous Solid Materials

2.1 INTRODUCTION

Drying of porous solids, such as fruits, vegetables, and woods, is a complex phenomenon performed by applying considerable amounts of thermal energy to dried materials. This process comprises two distinct mechanisms: (i) internal simultaneous heat transfer (from the drying air and through the dried material) and moisture (water) movement (inside the material via liquid water capillary flow and vapor diffusion); and (ii) simultaneous heat transfer and moisture (water) vaporization (forced evaporation) from the material surface to the environmental drying medium (usually, air).

In batch dryers, both the rate of moisture movement within the dried materials (i.e., the internal drying process) and the rate of moisture vaporization from the materials' surfaces (i.e., the external drying process) determine the drying durations. These rates mainly depend on (i) total amount of heat supplied per unit of time; (ii) material structure and transport properties; (iii) drying air dry- and wet-bulb temperatures, relative humidity, flow rates (or velocities), and airflow directions; and (iv) heat and mass transfer coefficients through the laminar, transient, and turbulent boundary layers.

During drying, the heat and moisture transfers are accompanied by temperature and moisture content variations, chemical and biological activities, and various product mechanical changes (Figure 2.1).

Various theoretical, semi-theoretical, and semiempirical models, such as Newton, Page, Logarithmic, and Exponential, are used to describe the drying mechanisms of porous solid material (www.academia.edu, accessed July 12, 2025) (Table 2.1).

2.2 INTERNAL COUPLED HEAT AND MOISTURE TRANSFERS

Heat transfer and moisture (mass) movement inside dried materials are strongly dependent on ambient temperature. They are, respectively, based on a coupling between Fourier's law (which establishes the temporal and spatial distribution of temperature) and Fick's laws dealing specifically with the moisture (water) diffusion (Luikov 1966; Whitaker 1998).

2.2.1 Internal Heat Transfer

Heat conduction without internal heat generation consists of heat transfer from the hotter ends to the colder ends of porous solid materials by vibrations and collisions of free electrons and molecules.

DOI: 10.1201/9781003618386-2

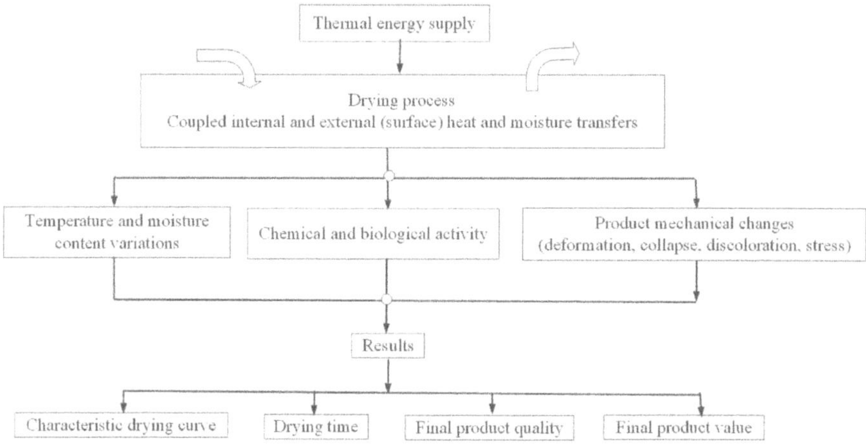

FIGURE 2.1 Main thermodynamic processes and results involved in drying of porous solid materials.

In a steady-state conduction heat transfer (Jiji 2009; Nellis and Klein 2024), (i) the temperature distribution in solid does not change with time ($\partial T/\partial t = 0$); (ii) the amount of heat entering any region of solid body is equal to the amount of heat coming out ($E_{in} = E_{out}$); and (iii) the temperature difference that drives the process is constant.

In a non-steady-state conduction, (i) temperature at any place within the porous solid materials change with time ($\partial T/\partial t \neq 0$); (ii) if changes of external drying medium are too rapid, the internal temperature distribution remains in a transient state; and (iii) energy entering the solid is not equal to the energy leaving it ($E_{in} \neq E_{out}$).

In steady-state thermodynamic systems, heat transfer by conduction is expressed by Fourier's law that relates the heat flux in a given direction to the temperature gradient in that direction. In one-direction form, it can be expressed as follows (Figure 2.2):

$$\dot{q} = -k \cdot \left(dT/dy\right) \tag{2.1}$$

TABLE 2.1
Mathematical Models Describing Drying Mechanisms

Name	Equation
Newton	$MR = e^{-kt}$
Page	$MR = e^{-kt^n}$
Logarithm	$MR = a \cdot e^{-(kt)} + a_1$
Exponential	$MR = a \cdot e^{-kt} + (1-a) \cdot e^{-kat}$

Note: MR: moisture ratio; k, n, a, a_1: constants determined by multiple regression techniques.

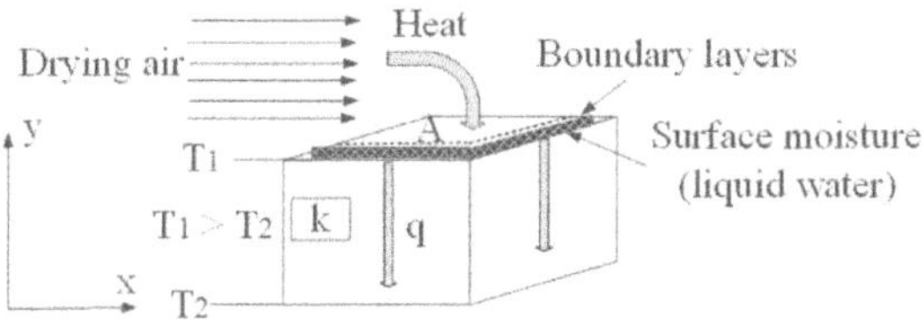

FIGURE 2.2 Schematic representation of Fourier's law of heat conduction.

where

$\dot{q}$ is the density of thermal power (W/m^2);
k is the material thermal conductivity ($W/m \cdot K$);
T is the material temperature (K);
y is the direction of heat transfer (m); and
dT/dy is the temperature gradient (acting as a driving force for internal heat transfer) (K/m).

Convective heat transfer at dried materials' surfaces is completed by internal conductive heat transfer (Figure 2.2) (Varnier et al. 2020):

$$\dot{q} = h_{conv}^{heat} \cdot (T_\infty - T_s) = k \cdot (dT/dy) \tag{2.2}$$

where

k is the thermal conductivity of porous solid material ($W/m \cdot K$);
T is the temperature (K);
y is the distance (m);
h_{conv}^{heat} is the air convective heat transfer coefficient ($W/m^2 \cdot K$);
T_∞ is the mean temperature of drying air (K); and
T_s is the mean temperature at the material surface (K).

According to Fourier's law of heat conduction, the rate of heat transfer through a material of surface A is proportional to the temperature negative gradient and the area (perpendicular to the temperature gradient) of the surface through which the heat flows:

$$\dot{Q}_{cond} = -k \cdot A \cdot (\Delta T / \Delta y) \tag{2.3}$$

where

$\dot{Q}_{cond}$ is the total conductive thermal power transferred (W);
k is the material thermal conductivity ($W/m \cdot K$);
T is the material temperature (K); and
y is the thickness of the material sample (m).

TABLE 2.2
Moisture (Mass) Transfer during Drying

	Conditions
Liquid diffusion	Due to capillary forces and moisture gradients Liquid vaporizes within the material
Vapor diffusion	Diffusion in pores filled with air
Hydrostatic pressure differences	Vaporization rate of interior moisture exceeds the rate of vapor being removed from the dried material

2.2.2 Internal Moisture Movement

During drying, moisture (water) diffuses (moves) from the centers of dried products to their surfaces. The overall rate (speed) and magnitude of moisture movement depend on materials' temperatures, relative dryness, and properties of drying medium (usually hot and dry air).

Moisture movement inside dried porous solid materials is driven by (Mujumdar 2000; Mujumdar and Huang 2007) (Table 2.2) (i) liquid diffusion due to capillary forces that allow the liquid transfer from zones with large moisture contents to zones with low moisture contents; (ii) diffusion of water vapor; (iii) hydrostatic pressure differences; and (iv) vapor partial pressure gradients between the interior and outer surfaces of dried products, and external mass flow rates of drying air.

2.2.2.1 Fick's First Law

Fick's first law relates the moisture diffusive mass flux to the gradient of moisture content. It postulates that the moisture mass flux goes from the regions of high moisture contents to the regions of low moisture contents, and that its magnitude is proportional to the moisture content gradient (spatial derivative):

$$\dot{m}_{moisture} = -D_{diff} \cdot \left(d\varphi / dy\right) \tag{2.4}$$

where

$\dot{m}_{moisture}$ is the specific moisture (mass) diffusion flux (also called moisture diffusivity) that measures the amount of moisture that flows through an unit area during one unit time ($kg_{water}/m^2 \cdot s$);

D_{diff} is the moisture diffusion coefficient (or diffusivity) that depends on the material particle sizes, moisture content, and temperature (m^2/s);

$d\varphi / dy$ is the linear gradient of moisture (particles) concentration per unit volume. This is the driving force for the one-dimensional moisture diffusion that occurs in response to a moisture content gradient expressed as the change in moisture content due to a change in position ($kg_{water}/(m^3/m)$;

φ is the moisture concentration per unit volume (kg_{water}/m^3);

$d\varphi$ is the moisture concentration gradient ($kg_{water}/m^3/m$);

y is the moisture position (m); and

dy is the change in moisture concentration position (m).

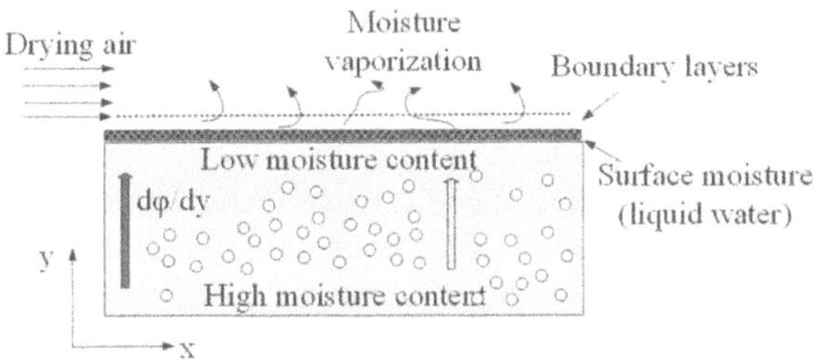

FIGURE 2.3 Schematic of linear gradient of moisture content per unit volume and density of moisture mass diffusion flux. φ, moisture concentration per unit volume; $\dot{m}_{moisture}$, density of moisture (mass) diffusion flux. *Note:* Schematic not to scale; not all components shown.

The negative sign of Equation 2.4 indicates that moisture diffusion occurs in a direction opposite to that of the increasing of moisture (particles) concentration. That is, diffusion occurs in the direction of decreasing concentration of the diffusing moisture, and thus the moisture diffusion flux is a positive quantity (Figure 2.3).

2.2.2.2 Fick's Second Law

For one-dimensional porous solid dried materials, Fick's second law of diffusion describes how the moisture (particles) concentration changes over time within a given volume. It describes the time-dependent behavior of the moisture concentration profile during diffusion. Fick's second law states that the rate of change of moisture concentration with respect to time is proportional to the second derivative of moisture concentration with respect to position with the diffusion coefficient (D_{eff}) as a proportionality constant. This means that moisture particles tend to move from areas of higher concentration to areas of lower concentration, leading to moisture equalization of concentration over time. In other words, Fick's second law predicts the changes in moisture concentration with respect to time (Crank 1975).

2.3 EXTERNAL (SURFACE) COUPLED HEAT AND MOISTURE TRANSFERS

In drying processes of porous solid materials, heat is transferred from the drying air flow to the developing boundary layer along the material surface. Then, the water vaporizes and diffuses from the liquid film on the dried material surface to the drying forced air stream. To maintain constant drying rates, the water molecules inside the dried products must absorb additional heat from the exterior to increase their kinetic energy.

Heat is transferred into the material interior primarily by conduction which depends on the material structure and thermal conductivity. When sufficient thermal energy is provided to cause the vaporization (forced evaporation) of water, the vapor is transported from inside of the product to the product surface before it can be transported away by the convective hot drying air.

Vaporization (forced evaporation) of liquid moisture takes place from the exposed surface by absorbing enthalpy (latent heat) of evaporation. The liquid evaporates from the drying surface of the product because of the difference of partial vapor pressures between the drying air and the surface of the product. This process depends on the amount of energy supplied, drying air moisture (mass) transfer coefficient, temperature, flow rate (or velocity), and capacity to absorb humidity. To maintain a steady vaporization (forced evaporation) rate, the vapor partial pressure of the drying air must be reduced by either increasing its temperature or reducing its relative humidity. The drying air temperature must be always expressed in terms of both dry-bulb and wet-bulb temperatures, or of depression temperature.

Moisture (mass) transfer rate from the materials' wet surfaces to the drying air depends on the surface thin stagnant moisture (water) film and temperature. In turn, this vaporization (forced evaporation) process induces a transport of moisture (water) from the interior of dried materials to their external exchange surfaces.

2.3.1 Boundary Layers

Consider a steady-state, external, parallel flow at zero incidence of a moist air stream (i.e., mixture of dry air and water vapor) of velocity u_∞ (m/s), temperature T_∞ (°C), and moisture concentration (or absolute humidity) $C_\infty(\text{kg}_{water}/\text{kg}_{dry-air})$ over a continuous flat plate of length L, area A_s, and uniform surface temperature $T_s < T_\infty$ (Figure 2.4) (Minea 2019). These external thermal and hydraulic parameters, as well as the relative and absolute humidity of the drying air, control the boundary conditions and affect the rate of moisture vaporization (i.e., the material drying rate) (Keey et al. 2000).

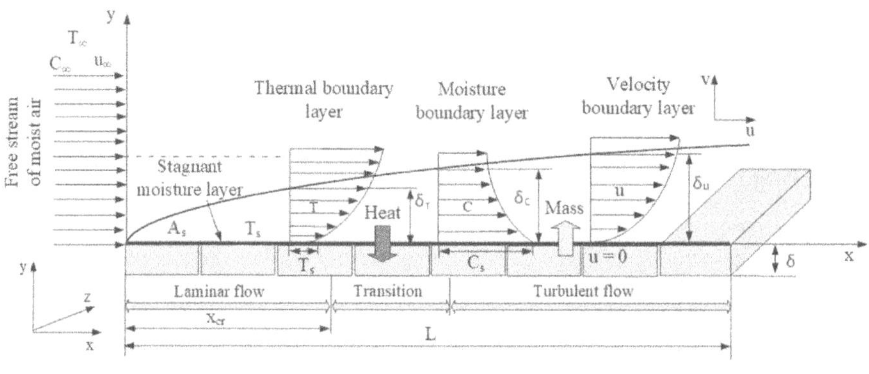

FIGURE 2.4 Schematic representation of thermal, moisture, and velocity boundary layers developed over a horizontal piece (specimen) of a dried porous solid material. A_s, surface area of dried piece; $C(\varphi)$, moisture concentration; δ, thickness of dried piece; L, width of dried piece; T_∞, air free-stream temperature; T_s, temperature of dried piece surface; u, velocity component of drying air in x direction; v, velocity component of drying air in y direction; x_{cr}, critical length; δ_C, thickness of moisture boundary layer; δ_T, thickness of thermal boundary layer; δ_u, thickness of velocity boundary layer. *Note:* Drawing not to scale; not all details and components shown.

The airflow over the flat plate initially starts in laminar region, but at some distance from the leading edge, small disturbances amplify, and the transition to turbulent flow begins to occur. Fluid fluctuations begin to develop in the transition region, and the boundary layer eventually becomes completely turbulent. In the fully turbulent region, fluid motion is highly irregular and is characterized by velocity fluctuations.

As can be seen in Figure 2.4 (which schematically represents the approximative configurations of drying air thermal, moisture, and velocity boundary layers at the surfaces of typical dried materials such as those of sliced fruits and vegetables and wood lumbers), the laminar flow begins to develop downstream the board leading edge ($x = 0$), the flow characterized by highly ordered air movement and, along a streamline, variable velocity components in both directions x and y (u and v, respectively).

At a certain location (x_{cr}), the transition to turbulent flow regime occurs (Incropera and DeWitt 2002).

2.3.2 Surface Heat Transfer

Heat transfer from the drying medium (usually air) to the surfaces of dried porous solid materials is first achieved by forced convection and then by conduction through to the interior of materials (Incropera and DeWitt 2002).

For pieces of dried materials in the form of flat plates (Figure 2.4), Nusselt dimensionless number (which provides a measure of the convective heat transfer) is one universal function of x^*, Re_L, and Pr:

$$Nu = f\left(x^*,\ Re_L, Pr\right) \tag{2.5}$$

where

Nu is the Nusselt dimensionless number (–);
$x^* = x/L$ is a dimensionless independent variable (–);
x is the distance downstream from the edge of the plate (m);
L is the length of the flat plate (m);
Re_L is the Reynolds number (–); and
Pr is the Prandtl number (–).

Over the entire lengths of material flat plates ($L \gg x_{cr}$), the average Nusselt number is defined as follows:

$$\overline{Nu}_L = (\overline{h}_{heat} \cdot L)/k_{air} = 0.037 Re_L^{4/5} Pr^{1/3} \tag{2.6}$$

where

$\overline{Nu}_L$ is the average Nusselt number (–);
$\overline{h}_{heat}$ is the average heat transfer coefficient ($W/m^2 \cdot K$);
L is the flat plate total length (m);
k_{air} is the air thermal conductivity ($W/m \cdot K$);
Re_L is the Reynolds number (–); and
Pr is the Prandtl number (–).

Equation 2.6 requires evaluation of fluid (drying air) properties at the reference (film) temperature:

$$T_{ref} = \left(\left(\bar{T}_\infty + \bar{T}_s\right)/2\right) \tag{2.7}$$

where
T_{ref} is the reference temperature (°C);
$\bar{T}_\infty$ is the drying air average temperature at infinite distance (°C); and
$\bar{T}_s$ in the average temperature of material surface (°C).

Under the assumption stating that spatial and temporal variations along the surface are not accounted for, the average heat convection coefficient ($\bar{h}_{heat}$) can be determined as follows:

$$\bar{h}_{heat} = (\overline{Nu}_L \cdot k_{air})/L \tag{2.8}$$

where
$\bar{h}_{heat}$ is the average heat transfer coefficient ($\text{W/m}^2 \cdot \text{K}$);
$\overline{Nu}_L$ is the average Nusselt number (–);
k_{air} is the air thermal conductivity ($\text{W/m} \cdot \text{K}$); and
L is the flat plate total length (m).

Typical values of $\bar{h}_{heat}$ may vary from $10\,\text{W/m}^2 \cdot \text{K}$ (forced convection with low speed flow over flat plates) to $100\,\text{W/m}^2 \cdot \text{K}$(forced convection with airflow velocities greater than $5\,\text{m/s}$).

The density of the heat flux over unit area of a heat transfer surface can then be computed with Newton's law as follows:

$$\dot{q} = \bar{h}_{heat} \cdot \left(\bar{T}_\infty - \bar{T}_s\right) \tag{2.9}$$

where
$\dot{q}$ is the density of the heat flux (W/m^2);
$\bar{h}_{heat}$ is the heat transfer coefficient, characteristic of the given airflow and heat transfer surface $\left(\text{W/m}^2 \cdot \text{K}\right)$;
$\bar{T}_\infty$ is the bulk (average) air temperature (°C); and
$\bar{T}_s$ is the material surface average temperature (°C).

Since at normal drying rates, the effect of the moisture vapor leaving the material surface on the net heat transfer is very small, it may be ignored.

For a given velocity and temperature of the drying air flowing over a horizontal flat plate of area $A_{surface}$ and uniform temperature $\bar{T}_\infty > \bar{T}_s$, the total heat transfer rate for the entire surface may be expressed as follows:

$$\dot{Q} = \bar{h}_{heat} \cdot A_s \cdot \left(\bar{T}_\infty - \bar{T}_s\right) \tag{2.10}$$

where

$\dot{Q}$ is the total rate of heat transfer to the dried material surfaces (W);

$\bar{h}_{heat}$ is the average convective heat transfer coefficient at the flat plate–drying air interface ($W/m^2 \cdot K$);

A_s is the total surface area of dried material (m^2);

$\bar{T}_\infty$ is the surface temperature of drying air far from the surface of dried material (°C); and

$\bar{T}_s$ is the temperature at the flat plate–drying air interface (°C).

2.3.3 Surface Moisture Vaporization (Forced Evaporation)

In drying chambers, forced-circulated drying air supplies sensible heat and enthalpy (latent heat) of vaporization through the thermal boundary layer to the moisture existing on the surfaces of materials to be dried. Heat is the source from which the water molecules acquire the kinetic energy necessary for moisture vaporization. The drying air is, thus, cooled while the moisture vaporizes from the materials' surfaces. The water vapor removed by the airflow dissipates into the bulk of the drying agent (air) and carried away.

The removal of moisture as vapor from the materials' surfaces by vaporization (forced evaporation) depends on the air temperature, relative humidity, flow rate (or velocity), and areas of exposed surfaces.

At the beginning of drying processes, the relative humidity of the moisture at the materials' surfaces is practically 100%, and the heat supplied is entirely used for moisture vaporization. The temperature of the liquid film and the adjacent solid surface is expressed as the wet-bulb temperature of the moving drying air (assuming radiation and conduction effects are negligible).

During the drying process of porous solids, the dry-bulb temperature of the drying air is higher than that of material pieces. If both these temperatures are kept constant, the rate of vaporization will gradually decrease as the supply of moisture of the dried material diminishes and as the vapor pressure of the drying air increases. Therefore, to maintain a steady-state drying rate, the water molecules inside the dried material must acquire additional thermal energy (heat), or the vapor partial pressure of the drying chamber atmosphere must be reduced. This is achieved by either increasing the air dry-bulb temperature (i.e., by supplying more heat) or reducing the drying air relative humidity. This suggests that the rate of moisture movement, whether into or out of the material pieces, is in large part dependent upon temperature and the relative dryness of the drying air and the dried material.

In convective heat transfer, the ratio of momentum diffusivity to thermal diffusivity is designated as Prandtl number:

$$Pr = \frac{Momentum\ diffusivity}{Thermal\ diffusivity} = \nu / a = \bar{c}_p \cdot \mu / k \tag{2.11}$$

where

Pr is the Prandtl number (–);

ν is the kinematic viscosity (momentum diffusivity (m^2/s);

a is the thermal diffusivity (m^2/s);
$\bar{c}_p$ is the specific heat ($J/kg \cdot K$);
μ is the dynamic viscosity ($Pa \cdot s = N \cdot s/m^2$); and
k is the thermal conductivity ($W/m \cdot K$).

In convective moisture (mass) transfer, Prandtl number is replaced by the Schmidt dimensionless number defined as the ratio of momentum diffusivity to mass diffusivity:

$$Sc = \frac{Momentum\ diffusivity}{Mass\ diffusivity} = \nu / D_{eff} = \mu / \left(\rho \cdot D_{eff}\right) \tag{2.12}$$

where
Sc is the Schmidt number (–);
ν is the kinematic viscosity (momentum diffusivity (m^2/s);
D_{eff} is the effective moisture diffusivity (m^2/s);
μ is the dynamic viscosity ($Pa \cdot s = N \cdot s/m^2$); and
ρ is the density (kg/m^3).

On the other hand, the Sherwood number provides a measure of the convective moisture (mass) transfer occurring at the board's surface. It is, thus, to the moisture (concentration) boundary layer what the Nusselt number is to the thermal boundary layer. Over the entire length of a flat plate (L), the average Sherwood number depends only on the Reynolds number (Re_L) and the Schmidt number (Sc), and is defined as follows:

$$\overline{Sh} = f\left(x^*, Re_L, Sc\right) = (\bar{k}_{mass} \cdot L) / D_{eff} \tag{2.13}$$

where
$\overline{Sh}$ is the Sherwood number (–);
$x^* = x/L$ is a dimensionless independent variable (–);
Re_L is the Reynolds number (–);
Sc is the Schmidt number (–);
$\bar{k}_{mass}$ is the average moisture (mass) transfer coefficient (m/s);
L is the flat plate length (m); and
D_{eff} is the moisture (water vapor) effective diffusion coefficient (m^2/s).

For external forced mixed laminar and turbulent flow over the entire surface of the flat plate, the average Sherwood number can be calculated as follows (for T_{ref} and $0.6 \leq Sc \leq 50$) (Incropera and DeWitt 2002):

$$\overline{Sh_L} = (0.037 * Re_L^{4/5} - 871) \cdot Sc^{1/3} \tag{2.14}$$

Thus, in addition to the air velocity, the convective moisture (mass) transfer coefficient is dependent on the length and the geometry of the vaporization surface, and the direction of the flow.

Equations 2.13 and 2.14 require evaluation of the fluid properties at the reference (film) temperature:

$$T_{ref} = (T_\infty + T_s)/2 \tag{2.15}$$

where

T_{ref} is the reference temperature (°C);

$\bar{T}_\infty$ is the surface temperature of drying air far from the surface of dried material (°C); and

$\bar{T}_s$ is the temperature at the flat plate–drying air interface (°C).

Total (convective) moisture (mass) transfer rate from a flat plate surface can be expressed as follows (Anderson 2014; Incropera and DeWitt 2002):

$$\dot{M}_{moisture} = \bar{k}_{mass} \cdot A_s \cdot M \cdot (\bar{\psi}_s - \bar{\psi}_\infty) \tag{2.16}$$

where

$\dot{M}_{moisture}$ is the total convective moisture (mass) transfer rate (kg_{water}/s);

$\bar{k}_{mass}$ is the average moisture (mass) transfer coefficient (m/s);

A_s is the moisture mass transfer surface area (m^2);

M is the moisture (water vapor) molar mass ($kg_{water}/kmol$);

$\bar{\psi}_s$ is the average moisture molar concentration at flat plate surface ($kmol/m^3$); and

$\bar{\psi}_\infty$ is the average moisture molar concentration far from the flat plate surface ($kmol/m^3$).

From Equation 2.16, it can be seen that to calculate the total convective moisture (mass) transfer rate from a material flat plate surface, it is necessary to determine the value of $\bar{\psi}_s$ and $\bar{\psi}_\infty$ by assuming thermodynamic equilibrium at the interface between the the material solid surface and the drying air. Such an equilibrium implies that (i) temperature of the moisture vapor at the interface is equal to the surface temperature (T_s); and (ii) moisture vapor is in a saturated state in which case thermodynamic tables for water may be used to obtain its density from knowledge of surface mean temperature.

2.4 TYPICAL DRYING PERIODS

In laboratory- and industrial-scale dryers, different value-added products (such as fruits, vegetables, and woods) are dried from the given initial moisture contents to the specified (suitable) final moisture contents.

Drying by forced hot air convection of porous solid materials is performed by passing warm drying air over each product. Then, since the saturation vapor pressure of the moisture over the solid product is less than that of the drying air, the moisture is transferred to the drying air and carried away. This process is mainly governed by the capacity of the surrounding forced drying air to absorb moisture from the surface of dried products.

At the beginning of solid porous materials' drying processes, the inner water liquid and vapor phases have uniform temperatures. When water evaporates, the latent heat of vaporization cools the surfaces of porous products, leading to a decrease of the saturated vapor pressure at the surface and, subsequently, to lower drying rates. The surface temperatures will decrease until an equilibrium between the latent heat of vaporization and the heat supplied is established. The incoming heat fluxes are composed of the heat transfer from the drying air forced flow due to thermal convection and thermal conduction through the porous materials. Presumably, the heat of vaporization is withdrawn partly from the water phase, which is assumed to be nearly in equilibrium with the solid material, and from the surrounding forced drying air.

Representative drying curves, which are determined at specific temperatures, velocities, and pressures of the drying air, exhibit successive drying periods from the beginning to the end of drying cycles. Three major drying periods can be observed on drying curves (Figure 2.5a and b): (i) warming-up period, during which the product is heated up; (ii) constant rate period, in which moisture is relatively easy to remove; and (iii) first and second falling rate periods, in which moisture is bound or held within the solid matrix.

2.4.1 Initial Warming-Up Period

Material drying behavior depends on its porosity, homogeneity, and hygroscopic properties. It can be characterized by drying curves that plot the drying rates as functions of time (Figure 2.5a) or drying curves that plot the drying rates as functions of moisture contents (Figure 2.5b).

During the initial drying period ($A-B$) (Figure 2.5a and b), the dried material is heated up slowly from the inlet condition to process condition. At this phase, the drying rate increases dramatically as heat supplied promotes greater rate of evaporation of the free moisture content on the material surface. During the warming-up period, the material stack(s) is (are) heated by air forced convection and conduction until the desired dry-bulb and/or wet-bulb temperatures are reached, prior to starting the first step of drying processes. The stacks of dried materials are heated, generally between 1 hour and 3 hours in the case of fruits and vegetables, and between 3 hours and 8 hours in the case of woods, depending on lumber boards' structures and initial temperatures. As heat sources for warming-up periods are generally used oil and natural gas (to produce steam at pressures up to 207kPa and 134°C saturated temperature, or even hot water), or electrical boilers (to produce hot air).

2.4.2 Constant Drying Rate Period

At the end of initial, relatively short, warming-up thermal regime, the materials' surfaces are completely covered by a thin layer of liquid water supplied from the material interior to the material surface by capillary flow. The vapor partial pressure at the material surface remains equal with the partial pressure of saturated vapor, and heat and moisture mass transfers occur simultaneously at the product surface.

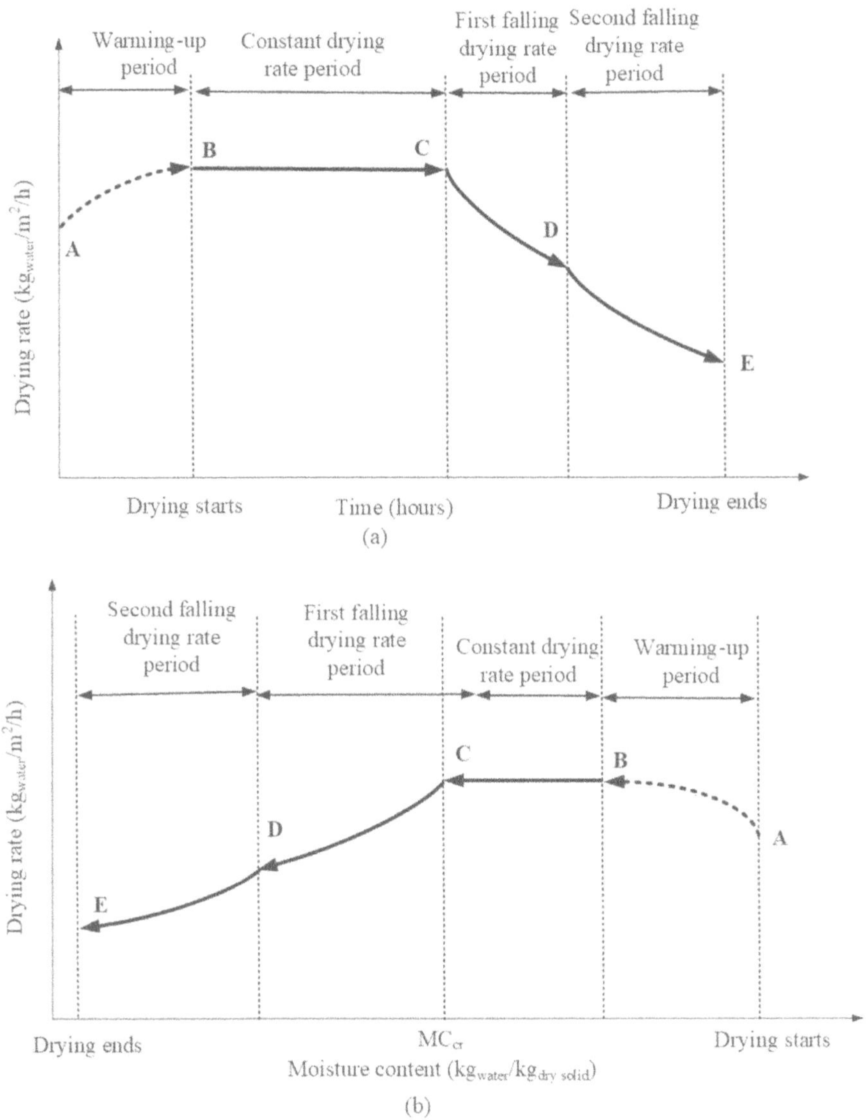

FIGURE 2.5 Typical drying periods of porous solid materials (a) versus drying time; (b) versus moisture content. MC_{cr}, critical moisture content. *Note:* Schematic not to scale.

After the warming-up period, the constant drying rate period begins and progresses by increasing the dry-bulb temperature or by decreasing the relative humidity of the drying medium (air) (process $B-C$) (Figure 2.5a and b). Constant drying rate is defined as the period of drying where moisture removal occurs at the surface by evaporation and the internal moisture transfer is high (sufficient) enough to maintain the saturated surface; thus, the evaporation rate remains constant (Figure 2.6).

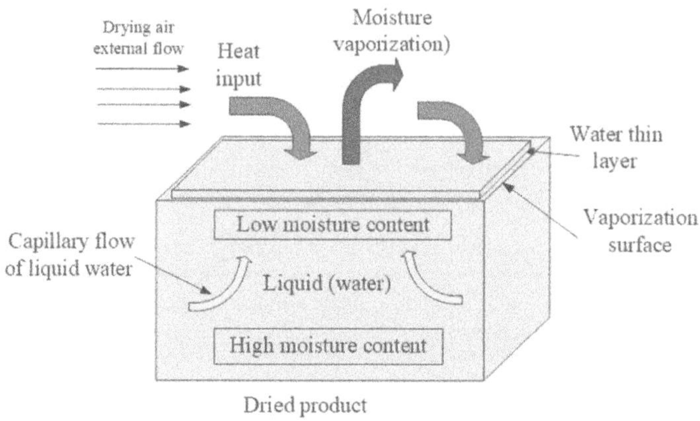

FIGURE 2.6 Schematic of internal and external coupled heat and moisture transfers during constant rate drying period content. *Note:* Schematic not to scale; not all components and processes shown.

During the constant rate period (Figure 2.5a and b), (i) materials' exposed surfaces are supplied with liquid water coming from the inside of dried materials by capillary action and other (inertial, viscous, and gravitational) driving forces; they remain saturated with liquid water since the rate of water movement is equal to the rate of moisture vaporization; (ii) internal moisture movement to the surface is governed by the material thickness and intrinsic fluid transport properties; (iii) drying continues by movement of water vapor from materials' saturated surfaces through stagnant air films into the mainstream of drying air; (iv) moisture content strongly decreases near the material surfaces; (v) vaporization of moisture (water) at the materials' surfaces is continuous; (vi) drying rate is high and relatively constant and depends only on the external conditions (drying air temperature, relative humidity, velocity, and flow configuration); (vii) moisture adhering to the products' surfaces is removed by the vaporization which is controlled by a convective process; moisture (water) vaporized from the material surface is compensated by the water diffused out from the interior to the surface of material; (viii) external heat provided to the product is entirely used as water latent heat of vaporization; (ix) velocity of drying air, together with the material geometry, define the heat and mass transfer coefficients, and, consequently, the drying rate depends on external conditions only; in other words, drying in the constant rate period is a surface-based process governed by external conditions such as temperature difference between the drying air and material wet surface, area exposed to the drying air, and external heat and mass transfer coefficients; (x) at low moisture contents, the internal rate of moisture movement (transport) is limited by the ability of the water to diffuse through the material to its surface; (xi) the rate of moisture removal during the constant rate period is limited by the ability of the circulating drying air to remove the water from the material surface; (xii) drying rate is dependent on the rate of heat transfer to the dried material and is equal to the rate of surface moisture (water) vaporization; (xiii) constant drying rate period lasts as long as the surface is supplied with liquid; its duration depends strongly on the drying conditions (e.g., drying air velocity and its thermophysical properties), and proceeds until the surface film of liquid is entirely vaporized; (xiv) constant rate drying period ends when the migration rate of water from the interior of the surface becomes less than the rate of evaporation from the surface; in other words, constant drying rate period ends once the vapor pressure is no longer equal to the saturated vapor pressure; (xv) at this moment, the drying rate starts to decrease and the falling rate period begins; and (xvi) at the end of constant drying rate period (i.e., the critical moisture content MC_{cr} is reached), the material surface can be assumed dry; from this point onward, the drying rate decreases, and the product temperature gradually increases; the driving forces across the boundary layers are progressively reduced, which signify the commencement of the first falling (decreasing) rate drying period.

2.4.3 Falling Drying Rate Periods

Falling rate period can be separated into two regions (Figure 2.5a and b): (i) first falling drying rate ($C-D$); and (ii) second, falling drying rate ($D-E$). It consists of two processes (Figure 2.7): (i) movement of moisture within the material to the surface by capillary flow of liquid water; and (ii) removal of moisture from the surface by vaporization.

During falling rate drying periods, (i) surfaces of dried materials are not saturated; (ii) the rate of moisture movement from the interior toward the materials' surfaces is less than the rate of vaporization from the surfaces; (iii) materials' surface temperatures approach the dry-bulb temperature of the drying air; (iv) consequently, the moisture vaporization is an internally controlled mechanism; and (v) drying rates slowly decrease until the material comes to equilibrium with the drying air.

2.4.3.1 First Falling Drying Rate Period

The moisture content at the end of the constant rate period (point C) is the critical moisture content (MC_{cr}). At this point, the surface of the solid is no longer saturated, and the rate of drying decreases with the decrease in moisture content. At point C,

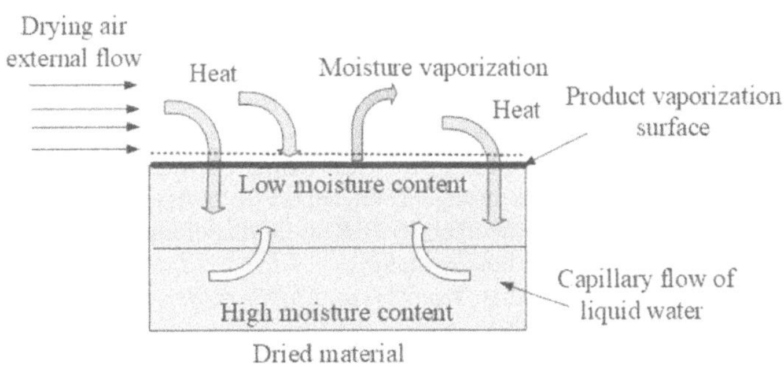

FIGURE 2.7 Schematic of internal and external coupled heat and moisture transfers during falling rate drying periods. *Note*: Schematic not to scale; not all components and processes shown.

the surface moisture film evaporates fully, and with the further decrease in moisture content, the drying rate is controlled by the rate of moisture movement through the solid.

During the first falling rate period ($C-D$ in Figure 2.5a and b), (i) interior moisture diffuses out to the material surface in order to maintain the moisture vaporization; (ii) thus, the rate of evaporation is controlled by the rate of internal moisture transfer; this process will continue until the wetted surface of the drying material is completely dry; (iii) drying rate decreases rapidly as the surface temperature rises, and moisture content decreases due to the additional internal resistance for moisture transfer; and (iv) first falling rate period proceeds until the surface film of liquid is entirely evaporated; this means that the material's surface is dried out.

2.4.3.2 Second Falling Drying Rate Period

On further drying, the moisture moves through the solid because of moisture gradients between the material deeper parts and the surface.

The second falling rate period (process $D-E$ in Figure 2.5a and b) begins when the partial pressure of water throughout the material is below the saturation point. The heat flux from the hot drying air to the dried material is very low because of small temperature gradients between the hot drying air and the material surface, which is now close to the hot drying air temperature.

During the second falling rate period, the drying rate is largely independent of conditions outside the solid and the moisture (mass) transfer is due to combinations of liquid and vapor diffusion, and capillary moisture movement. The zone of vaporization recedes slowly below the surface and the remaining water inside the drying material vaporizes by the heat conducted through the material and diffuses into the drying air stream. Finally, the drying process terminates when free moisture content of the product approaches zero and equilibrium moisture content is achieved. In other words, at the end of the second falling drying period, the moisture content of dried material reaches the minimum value.

REFERENCES

Anderson, J.O. 2014. Energy and resource efficiency in convective drying systems in the process industry. Doctoral thesis, Lulea University of Technology, Sweden.

Crank, J. 1975. *The Mathematics of Diffusion*, Clarendon Press, Oxford, pp. 28–43.

Incropera, F.P., D.P. DeWitt. 2002. *Fundamentals of Heat and Mass Transfer*, 5th edition, John Willey & Sons, Inc., Hoboken, NJ.

Jiji, L.M. 2009. One-dimensional steady-state conduction. In: *Heat Conduction*, Springer, Berlin, Germany (https://link.springer.com/chapter/10, accessed February 3, 2024).

Keey, R.B., T.A.G. Langrish, J.C.F. Walker. 2000. Kiln-Drying of Lumber, Springer Series in Wood Science, Springer, Berlin, Germany.

Luikov, A.V. 1966. *Heat and Mass Transfer in Capillary-Porous Bodies*, 1st edition, Pergamon Press, New York, NY.

Minea, V. 2019. *Industrial Heat Pump-Assisted Wood Drying*, CRC Press, Boca Raton, FL.

Mujumdar, A.S., S. Devahastin (Ed.). 2000. *Mujumdar's Practical Guide to Industrial Drying, Principles, Equipment and New Developments*, Thananuch Business Ltd. Publication, Bangkok, Thailand.

Mujumdar, A.S., L.X. Huang. 2007. Global R&D needs in drying. *Drying Technology* 25:647–658.

Nellis, G., S. Klein. 2024. *Heat Transfer*, Cambridge University Press (https://assets.cambridge.org/, accessed February 3, 2024).

Varnier, M., N. Sauvat, L. Ulmet, C. Montero, F. Dubois. 2020. Influence of temperature in a mass transfer simulation: Application to wood. *Wood Science and Technology* 54(4):943–962.

Whitaker, S. 1998. Coupled transport in multiphase systems: A theory of drying. *Advances in Heat Transfer* 31:1–104.

3 Drying of Fruits and Vegetables

3.1 INTRODUCTION

Fruits and vegetables are perishable agricultural products prone to degradation. Their relative short shelf life can be extended, and at the same time the nutritional values of vitamins and minerals and natural flavor can be maintained by the following methods (Okos et al. 1992; Sagar and Kumar 2010): (i) refrigeration or freezing; (ii) removal of oxygen from packaging containers; (iii) processing into juices, jams, and jelly; (iv) canning, fermentation, and pickling; and (v) natural or mechanical drying.

All these procedures prevent the development of microorganisms such as bacteria and fungi, minimize biochemical, chemical, and microbiological deterioration, and reduce the mass and volume of dried products, thus facilitating packaging, handling, storage, and transportation requirements.

Drying of fruits and vegetables is an energy-intensive operation that comprises heat transfer between the surrounding media and surfaces of dried materials, simultaneous heat and moisture (mass) transfers within the dried materials, evaporation (vaporization) of moisture (water) at the surfaces of dried materials, and, finally, extraction of evaporated moisture from drying enclosures.

3.2 PHYTOCHEMICALS AND BIOACTIVITY OF FRUITS

Fruits are important sources of fibers, vitamins, phytochemicals (as flavonoids and phenolic acids), and antioxidants (Miller and Rice-Evans 1997; Chan et al. 2009) (Table 3.1).

Depending on fruit drying method used (Bai et al. 2013; Biesaga 2011; Nunes et al. 2016; Vinson et al. 2001), (i) additional phytochemicals may be released; (ii) dehydration can improve the fruit natural antioxidant properties; and (iii) at temperatures greater than 70°C, polyphenol oxidase in apples, avocados, grapes, pears, and plums could be inactivated.

3.3 METHODS FOR DRYING FRUITS AND VEGETABLES

Drying of fruits and vegetables can either be performed by solar (sun), vacuum (suitable for heat-sensitive materials), microwave (that can significantly reduce the drying time), freeze (that can maintain the microstructure of porous solid materials), hot air convective (batch or tray) methods, and more advanced technologies as pure heat pump-assisted drying or (solar, radio frequency, infrared) combined heat pump-assisted techniques (that aim to achieve more gentle drying processes and reduce energy consumptions) (Figure 3.1).

DOI: 10.1201/9781003618386-3

TABLE 3.1
Phytochemical and Bioactivity of Some Tropical Fruits

Fruit	Phytochemicals	Bioactivity
Dates	Carotenoids; flavonoids; phenolic acids; tannins	Antioxidant; anti-inflammatory
Durian	Flavonoids; tannins	Antioxidant
Guava	Flavonoids; terpenoids	Antioxidant; anti-inflammatory
Jackfruit	Carotenoids; flavonoids; fatty acids	Antibacterial; anti-inflammatory
Mango	Flavonoids; phenolic acids; tannin	Antioxidant
Papaya	Carotenoids; flavonoids; phenolic acids;	Antibacterial
Pineapple	Carotenoids; flavonoids; terpenoids	Antioxidant
Figs	Flavonoids; phenolic acids; anthocyanins	Antioxidant; antispasmodic

Selection of fruit and vegetable drying methods is generally based on product properties, such as (i) form (e.g., chip, cube, bar, or powder); (ii) cell structure; (iii) bioactive compounds; (iv) sugar content; and (v) initial and final required moisture contents.

Inappropriate drying methods of fruits and vegetables and inadequate operating conditions may result in unfavorable characteristics and properties of dried products such as (i) collapsed cell structures; (ii) stickiness; (iii) nutrient and bioactive substance degradation; and (iv) undesirable *external* (such as size, physical appearance and shape, color/discoloration, browning due to caramelization, consistency, texture) and internal (e.g., loss in flavor aroma and nutritive values) changes.

3.3.1 Solar (Sun) Drying

The daily world-average solar radiation (insolation) on horizontal surfaces is of about $3.82\,kWh/m^2$ with values in tropical countries being higher (up to $7.15\ kWh/m^2$).

Historically, the sun (and wind) drying method was used to remove moisture (water) from fruits, vegetables (and woods), and many other agricultural and biological products (e.g., meats, grains, and herbs).

Solar-assisted dryers present attractions in tropical or subtropical climates and remote locations spread over latitudes from 0° to 50°. In such climates, the dryers can easily exceed the average ambient temperature by 17°C with maximum differences of 22–33°C.

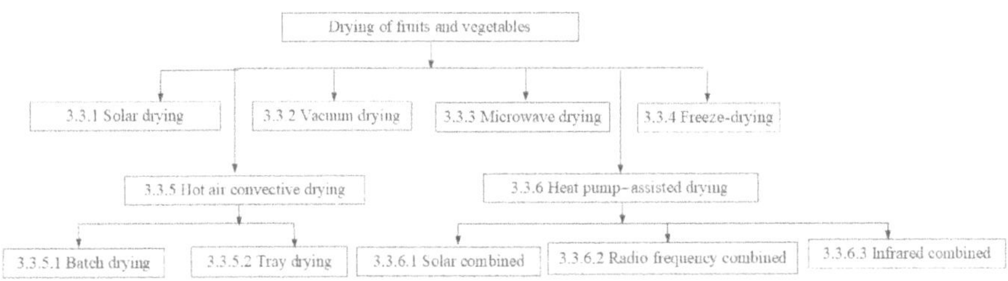

FIGURE 3.1 Conventional, advanced, and combined techniques for drying fruits, vegetables, and other agricultural products.

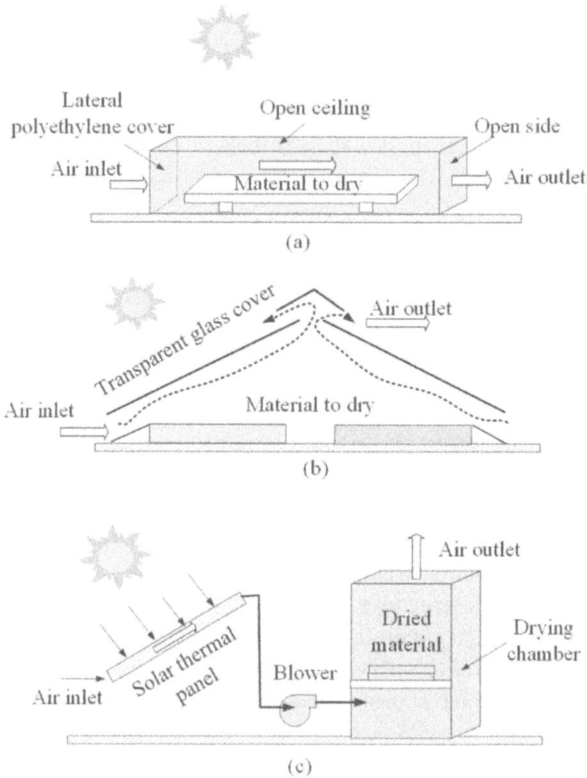

FIGURE 3.2 Types of solar (sun) dryers. (a) Direct (open). (b) Indirect (closed) (under transparent glass covers). (iii) Split (with solar thermal energy captors and forced drying air).

Traditional solar drying systems of fruits and vegetables are (i) direct (open) solar (sun) dryers (where the materials are spread in thin layers on large outdoor flat surfaces (Figure 3.2a); (ii) indirect (closed) (where the products are placed under transparent plastic or fiberglass covers and the airflow is driven by natural or forced convection using small wind turbines (Figure 3.2b); and (iii) split (where solar thermal energy captors heat the air that then flows through the dried materials by forced convection) (Figure 3.2c).

Compared to drying systems that use fossil and/or electric energy sources, solar drying of fruits and vegetables (usually at temperatures between 50°C and 60°C) is a more energy- efficient, low-cost, and environment-friendly technology.

The limitations of solar drying include (Tiris et al. 1994) (i) large dependence on the intermittent availability of solar radiation; (ii) relatively long drying times; (iii) labor-intensive and time-consuming operations; (iv) during open sun dying, materials that remain outdoor for long periods of time are subjected to climate changes, could be contaminated from air, precipitations, soil, dust, and rain, and losses could exist due to rodents, birds, and growth of microorganisms; (v) depending on weather conditions, over or under drying may occur; and (vi) product shrinkage could be important.

3.3.2 Vacuum Drying

Vacuum drying is used especially for drying relatively small quantities of heat-sensitive products as fruits and vegetables. It requires sealed enclosures where the air is evacuated to provide very low pressures, so that water can evaporate at much lower temperatures (Figure 3.3).

Vacuum dryers are usually heated by conduction via metal plates (that hold the dried products) using electrical resistances, saturated or superheated steam, or hot water as heating agents.

As can be seen in Figure 3.3, the enclosed vacuum drying chamber is connected to a mechanical vacuum pump (which helps to reduce the air pressures at between 9.593 kPa and 19.941 kPa in order to lower the corresponding boiling points of water at between 45°C and 60°C). The extracted water vapor (moisture) is then condensed over the fins of a (water or air) cooled condenser.

The following are the advantages of vacuum drying (Mitra et al. 2011): (i) moisture evaporates faster, accelerating the drying rates; (ii) can yield superior final quality (such as flavor, texture, color, and retention of bioactive ingredients); (iii) can reduce or prevent discolorations and oxidation of sensitive products that cannot be dried by convective hot air; and (iv) eliminates the heat losses through the exhaust air and, thus, is thermally more efficient.

Nevertheless, vacuum drying comes along with the following drawbacks (Reis 2014): (i) due to the need for construction of airtight vacuum chambers, the initial and operating costs are higher as compared to hot air convection dryers; (ii) vacuum pumps are relatively expensive; (iii) difficulties in getting adequate air circulation through the dried materials to obtain efficient heat transfers; (iv) lower efficiency when drying materials contain high amounts of moisture; (v) small production capacity; (vi) it is not convenient to access the drying materials during vacuum drying until the whole drying process has been completed; (vii) since there is no air inside the drying chamber, heat can only be supplied in the form of radiation or conduction (for example, heating up the plates which support the dried materials); and (viii) it is a drying method often used as a secondary drying technology aiming to reduce the moisture of hot air convective dried products from 20–25% (d.b.) to 1–3% (d.b.).

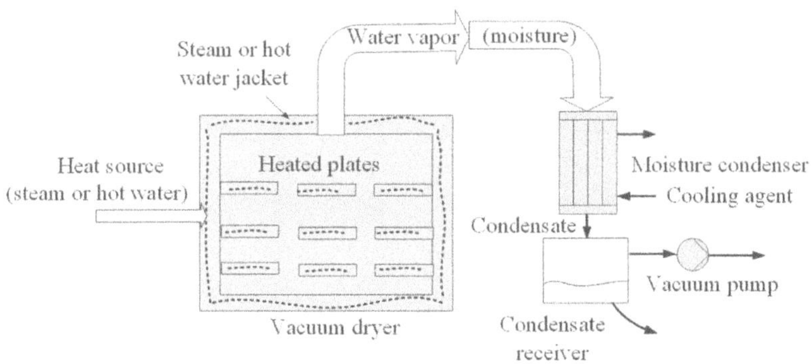

FIGURE 3.3 Schematic of a typical vacuum dryer. *Note:* Schematic not to scale; not all components shown.

3.3.3 Microwave Drying

Microwaves consist of electromagnetic radiation with frequencies ranging from approximately 3×10^8 Hz to 3×10^{11} Hz. To improve the quality of products as fruits and vegetables that contain dipolar or ionic substances, microwave drying could be an alternative drying method since heat is generated quickly inside the products and can penetrate deeply. Polarized ions accelerate and collide with each other, which cause the conversion of kinetic energy into thermal energy and eventually increases the temperature of the drying material. Consequently, water is evaporated rapidly (Mujumdar 2014).

Major components of typical microwave dryers are shown in Figure 3.4.

Microwave drying offers the following advantages (Maskan 2001; Rodríguez et al. 2005): (i) due to selective absorption of radiation by liquid water within fruits and vegetables, there will be only selective heating of product's interior without affecting the product's exterior; (ii) may give more uniform final moisture contents; and (iii) by right selection of microwave power input, microwave drying could save up to 50% of energy consumption with good retention of volatile organic compounds.

The following are the limitations of microwave drying: (i) inability to provide uniform energy distribution; (ii) may generate "hot" and "cold" spots inside the drying chamber; (iii) rapid rise of relative temperature gradient could have adverse effects on the quality and texture of dried products; (iv) may initiate further degradation of heat-sensitive bioactive ingredients (such as vitamin C and total phenolics contents); (v) expensive drying technology (especially for drying materials with high moisture contents); and (vi) care must be taken to shield the operators from the microwave radiation power density; radiation power density of $100\,\text{W/m}^2$ at a distance 0.05 m from the dryer is an acceptable limit.

3.3.4 Freeze-Drying

Freeze-drying (also known as cryodesiccation) involves the removal of water via sublimation directly from the solid state (ice crystals) to the water vapor by reducing

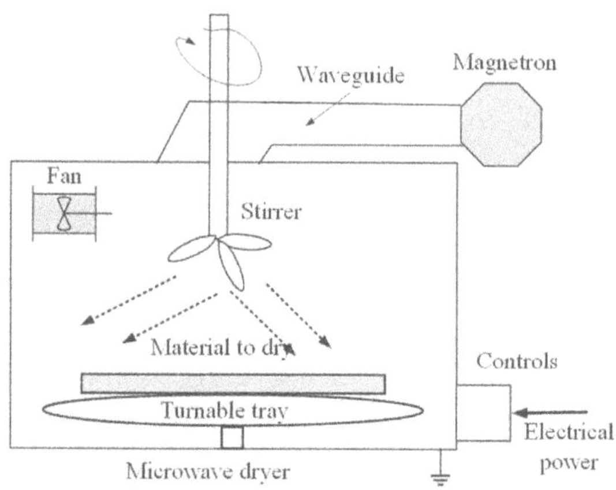

FIGURE 3.4 Major components of a microwave dryer. *Note:* Schematic not to scale; not all components shown.

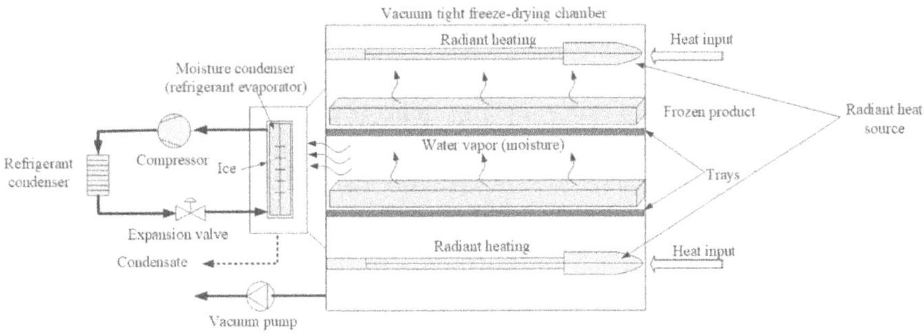

FIGURE 3.5 (a) Schematic of a freeze-dryer. *Note:* Schematic not to scale; not all components shown.

the pressure and temperature of surrounding air inside the drying chamber below the triple point of water.

This drying technique is traditionally used to dehydrate high-value pharmaceutical, biotechnological products (e.g., blood cells and plasma), some food products (e.g., spices and meats), and perishable fruits (e.g., guava and jackfruit).

Freeze-dryers include four basic components (Figure 3.5): (i) batch drying chambers provided with elements (trays, shelves, carts, etc.) for supporting 100–1,500 kg of materials being dried, and they are also capable of maintaining vacuum with minimal air leakage; (ii) vacuum pumps capable of producing deep vacuums; (iii) heat sources (e.g., hot water, or steam); (iv) moisture condensers (refrigerant evaporator) and their refrigeration systems; and (v) control and measurement instruments (pressures, temperatures, and, possibly, material mass).

The material to be freeze-dried is spread on aluminum trays. Heat is usually supplied by circulating a heating medium through the hollow shelves (floating shelfs) supporting the trays.

The water vapor is finally condensed over the finned tubes of moisture condensers (i.e., refrigerant evaporators).

In this type of drying, initial quick freezing (by the aid of refrigeration systems) of dried products is provided to obtain small ice crystals that enhance the thermal conductivity of dried materials by increasing their porosity.

The vacuum pumps remove the noncondensable gases, and, subsequently, reach the desirable vacuum level (e.g., less than 30 Pa). The heating sources (hot water or steam) provide the latent heat of sublimation once the frozen product is already at deep vacuum pressures. Heat conduction via the tray bottoms provides heat for sublimation.

The temperatures of the heating sources are determined by the ice transition temperature and the moisture content of the product. The moisture condensers (i.e., refrigerant evaporators) collect the water vapor released by the sublimation of ice crystals within the product. They must have sufficient heat transfer surfaces and cooling capacities to freeze all the moisture (vapor) generated during the sublimation process. The water vapor contacts the moisture condensers' surfaces and is changed into ice crystals releasing latent heat. Afterward, the ice crystals are removed from the system by moisture condensers. Typically, the operational temperature of moisture condensers (refrigeration evaporators) in commercial freeze-dryers varies from –50°C to –65°C.

The moisture condensers could be installed inside the drying chambers or in separate spaces flanged to drying chambers. Considering the very large volume of vapor to be moved by moisture condensers, the distance between trays and condensers should be kept as short as possible. In large units, two moisture condensers per vacuum drying chamber, working in alternation, could be used. While one condenser is in service, the other is isolated from the chamber and defrosted.

The following are the advantages of freeze-drying method: (Berk 2009) (i) original shapes, nutrient components, textural properties, freshness, color, flavor, aroma, and other physical and chemical properties could be kept for unlimited time; (ii) minimizes or avoids the material shrinkage; (iii) minimizes the movement of soluble solids within the dried products; (iv) fast rehydration and high retention of volatile compounds; (v) can prevent deterioration due to oxidation and enzymatic and nonenzymatic browning reactions of food products.

The following are the limitations of freeze-drying method: (i) used in food industries only for high-value food products; (ii) energy-intensive process; (iii) usually low temperatures and pressures prolong the total drying times and promote greater losses in volatile organic compounds; (iv) requires heavy duty freezing and vacuum pumping devices; consequently, its operating costs can be up to two to five times greater than other commercial drying methods; (v) levels of compounds in some fruits, such as soluble sugar and organic acids, may drop significantly during freeze-drying of some fruits (as mushroom slices); and (vi) majority of freeze-dryers are batch-type with very limited capacities.

3.3.5 Hot Air Convective Drying

The most common method to dry foodstuffs is the hot air convective drying that depends on drying conditions such as drying air temperature, relative humidity, velocity, and physical and chemical parameters of dried materials.

Hot air (forced) convective dryers are cabinet-type (Figure 3.6a) (partially not affected by frequent climate changes), tunnel-type with concurrent-flow devices (Figure 3.6b), or tunnel-type with countercurrent-flow (Figure 3.6c) devices.

In such driers, drying air heaters could be electrical coils or heat exchangers using hot water or steam as heat sources. Inside the drying chambers, electrical-driven blowers provide the circulation of hot convective air streams over the dried materials to facilitate the evaporation of moisture on the material surfaces.

Compared to solar dryers, convective hot air driers present the following advantages: (i) more efficient in providing faster and more uniform drying; (ii) ensure more hygienic dried products, and prolongs their shelf life by reducing the destructive potential of microbial and enzymatic activities; and (iii) provide relatively higher production rates.

However, hot air convective drying method has the following limitations (Lewicki 2006): (i) requires huge amount of energy, especially to raise up the system initial temperature and also to maintain relatively high temperatures inside the drying chambers and to provide sufficient high air velocities; (ii) large amounts of wasted thermal energy with the exhaust air (usually 35–45%); and (iii) at higher drying

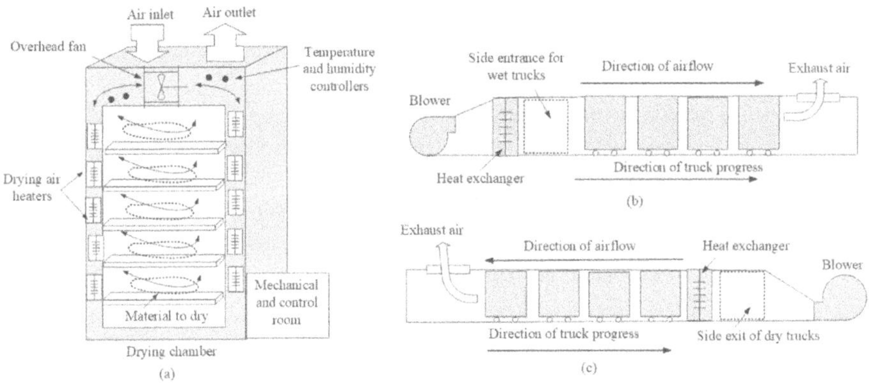

FIGURE 3.6 Schematic of hot air convective dryers. (a) Cabinet-type. (b) Tunnel-type with concurrent airflow. (c) Tunnel-type with countercurrent airflow. *Note:* Schematic not to scale; not all components shown.

temperatures, adverse effects (e.g., deterioration of bioactive ingredients, changes in nutritional properties, shrinkage, case hardening, and browning) on the quality of certain fruits (e.g., apples and bananas) can occur.

3.4 DRYING PARAMETERS OF FRUIT AND VEGETABLE

3.4.1 Moisture Content

Moisture contents of fresh fruits (e.g., bananas, mangoes, apples, grapes, pineapples, papaya, avocado, guava, durian, passion fruits, and peaches) and vegetables (e.g., carrots, ginger, pepper, and tomatoes) are relatively high ($\geq 80\%$, d.b.). This parameter can be calculated on dry basis (Equation 3.1) or on wet basis (Equation 3.2).

Dry basis (d.b.) moisture content of a material sample that can range from 0 to very large percentages is defined as the ratio of the mass of water to the mass of dry matter:

$$MC_{dry\,basis} = (m_{water} / m_{dry\ solid}) \times 100 \tag{3.1}$$

where

$MC_{dry\,basis}$ is the dry basis moisture content of product sample ($kg_{water} / kg_{dry\,solid}$) (%);
m_{water} is the mass of water contained in the product sample (kg_{water}); and
$m_{dry\,solid}$ is the mass of dry matter contained in the product sample ($kg_{dry\,solid}$).

Wet basis (w.b.) moisture content of a material sample that can range from 0% to 100% is defined as the ratio of the mass of water contained in a product sample to the total mass of product sample:

$$MC_{wet\,basis} = (m_{water} / m_{total\,solid}) \times 100 = \left[m_{water} / (m_{water} + m_{dry\,solid}) \right] \times 100 \tag{3.2}$$

where

$MC_{wet\ basis}$ is the wet basis moisture content of the product sample ($kg_{water}/kg_{total\ solid}$) (%);

m_{water} is the mass of water contained in the product sample (kg_{water}); and

$m_{total\ solid}$ is the total mass (water plus dry matter) of the product sample ($kg_{total\ solid}$)

If the initial dry basis moisture content of a product sample is known (based on product's experimental drying curves), the product moisture content at any time during the drying process can be determined as follows:

$$MC = MC_{in} \cdot e^{-k \cdot t} \tag{3.3}$$

where

MC is the dry basis moisture content at time t ($kg_{water}/kg_{dry\ solid}$) (%);

MC_{in} is the initial dry basis moisture content ($kg_{water}/kg_{dry\ solid}$) (%);

k is an experimental constant specific for each dried product (fruit or vegetable) (–); and

t is the drying time (s).

Based on Equation 3.3, the time necessary to reach the desired final (dry basis) moisture content can be calculated:

$$t = \ln\left(MC_{in}/MC\right)/k \tag{3.4}$$

where

t is the time necessary to reach the desired final (dry basis) moisture content (s);

MC_{in} is the initial dry basis moisture content ($kg_{water}/kg_{dry\ solid}$) (%);

MC is the dry basis moisture content at time t ($kg_{water}/kg_{dry\ solid}$) (%); and

k is an experimental constant specific for each dried product (fruit or vegetable) (–).

3.4.2 Moisture Ratio

Moisture contents of fruits and vegetables can be converted to moisture ratios using the following equation:

$$MR = \left[\left(MC_t - MC_{eq}\right)/\left(MC_{in} - MC_{eq}\right) \times 100\right] \tag{3.5}$$

where

MR is the moisture ratio (%);

MC_t is the dry basis moisture content at a given time t ($kg_{water}/kg_{dry\ solid}$) (%);

MC_{eq} is the equilibrium dry basis moisture content ($kg_{water}/kg_{dry\ solid}$) (%); and

MC_{in} is the initial dry basis moisture content ($kg_{water}/kg_{dry\ solid}$) (%).

Usually, the surrounding hot air is considered to be completely dry and, thus, M_{eq} can be assumed to be negligible. Therefore, Equation 3.5 can be reduced to the following:

$$MR = (M_t / M_{in}) \times 100 \tag{3.6}$$

where

MR is the moisture ratio (%);

M_t is the dry basis moisture content at a given time t ($kg_{water} / kg_{dry\,solid}$) (%); and

M_{in} is the dry basis moisture content at the beginning of drying cycle (kg_{water} / ($kg_{dry\,solid}$) (%).

3.4.3 Drying Rate

Drying rates of fruits and vegetables samples can be calculated as follows:

$$DR = \frac{\Delta(MC)}{\Delta t} = \left[\left(MC_{t+dt} - MC_t / \Delta t\right)\right] \tag{3.7}$$

where

DR is the drying rate (kg_{water} /s);

$\Delta(MC)$ is the variation of dry basis moisture content between two consecutive measurements ($kg_{water} / kg_{dry\,solid}$);

Δt is the drying time variation (s);

MC_{t+dt} is the dry basis moisture content at time $t + dt$ ($kg_{water} / kg_{dry\,solid}$); and

MC_t is the dry basis moisture content at time t ($kg_{water} / kg_{dry\,solid}$).

3.4.4 Effective Moisture Diffusivity

Movement of moisture (water) through dried fruits and vegetables can be explained by different mechanisms: (i) diffusion of liquid caused by concentration gradients; (ii) partial vapor pressure gradients; (iii) capillary forces; (iv) gravity forces; and (v) surface diffusion.

The moisture movement in fruits and vegetables is described by a parameter called effective moisture diffusivity (D_{eff}), an overall property of water mass transfer in dried materials. D_{eff} is typically modeled using Fick's second law under the assumption of negligible shrinkage. Its values vary due to factors such as material structure and properties and drying parameters (e.g., temperature). D_{eff} is also used to predict drying times.

Effective moisture diffusivity (D_{eff}) values for some fruits and vegetables are given in Table 3.2 (Mujumdar and Devahastin 2000).

3.4.5 Activation Energy

Activation energy is the minimum energy required to initiate the moisture movement from the interior to the surface of fruits and vegetables. It is a measure of dependence of moisture (water mass) migration process on drying temperature. The activation energy for most fruits and vegetables usually ranges from 12.7 kJ/mol to 110 kJ/mol.

TABLE 3.2
Effective Diffusivity of Some Food Products

Product	Moisture Content $kg_{water}/kg_{dry\,solid}$ (d.b.)	Temperature °C	Effective Moisture Diffusivity D_{eff} (m^2/s)
Apple	0.10–0.15	30–70	$1.0 \times 10^{-11} - 6.4 \times 10^{-9}$
Banana	0.01–3.5	20–40	$3.0 \times 10^{-13} - 2.1 \times 10^{-10}$
Carrot	0.01–5.00	30–70	$1.2 \times 10^{-9} - 5.9 \times 10^{-9}$
Pears	N/A	66	9.6×10^{-10}
Potato	N/A	54–68.8	$2.6 \times 10^{-11} - 6.4 \times 10^{-11}$

Note: N/A, not available.

3.5 INITIAL QUANTITY OF PRODUCT TO DRY

The initial quantities of fruits and vegetables to dry are always known since the physical dimensions (volumes, charging capacities) of hot air convective batch dryers coupled with mechanical vapor compression heat pumps are known.

Because heat pump-assisted dryers cannot accurately operate if the total sensible and latent heat extracted from the dried materials are not large enough to provide appropriate operating thermodynamic parameters to mechanical vapor compression heat pumps, it is mandatory to determine the quantity (mass, volume) of fruits and vegetables that have to be introduced inside the drying enclosures to provide sufficient latent heat input to the heat pumps' evaporators.

3.6 AVERAGE INITIAL MOISTURE CONTENTS OF PRODUCTS TO DRY

Table 3.3 shows average initial moisture contents of some fresh fruits and vegetables.

TABLE 3.3
Average Initial Moisture Contents of Some Fresh Fruits and Vegetables

Fruit/Vegetable	Moisture Content (%)
Guava	87.92 (w.b.)
Mango	87.20 (w.b.)
Pumpkin	84.57 (w.b.)
Tomato	90.61 (w.b.)
Asparagus beans	88.74 (w.b.)
Red cabbage	92.05 (w.b.)

Note: w.b., wet basis.

TABLE 3.4
Average Final Moisture Contents of Some Dried Fruits and Vegetables

Fruit/Vegetable	Moisture Content (%)
Guava	6.94 (w.b.)
Carrot	12.7 (w.b.)
Pears	6.2 (w.b.)
Mango	11.43 (w.b.)
Banana	9.95 (w.b.)
Plantain	9.97 (w.b.)
Mango	8.7 (w.b.)
Tomato	18.28 (w.b.)
Sweet potato	5.9 (w.b.)
Spinach	4.7 (w.b.)

Note: w.b., wet-basis.

3.7 AVERAGE FINAL MOISTURE CONTENT OF SOME DRIED FRUITS AND VEGETABLES

Generally, the average final moisture contents for many dried fruits and vegetables should be about 20% (w.b.) or below (Table 3.4).

3.8 TYPICAL DRYING CURVES

If the initial moisture contents (MC_{in}) (d.b.) of dried fruits and vegetables are known, Equation 3.3 allows determining the dry basis moisture contents at any time t. This section shows typical (representative) drying curves of some fruits (bananas, mangos, pineapples, kiwifruits, and apples) and vegetables (carrots and green peppers) (see Figures 3.7–3.13).

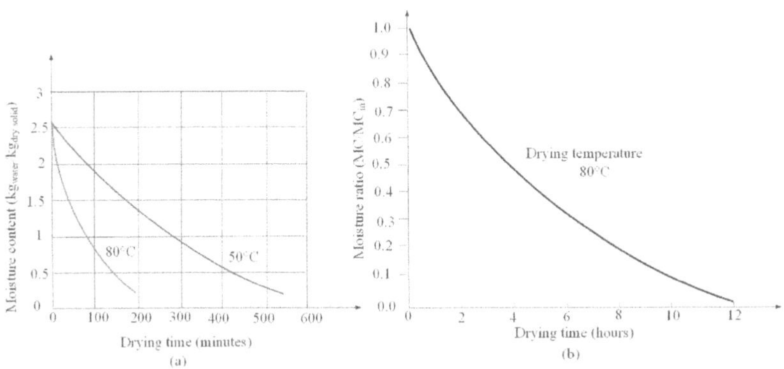

FIGURE 3.7 (a) Variation of moisture content of banana slices versus drying time for two extreme temperatures. (Redrawn from Doymaz 2010.) (b) Variation of moisture ratio versus time for banana dried slices. (Redrawn from Mercer 2014.)

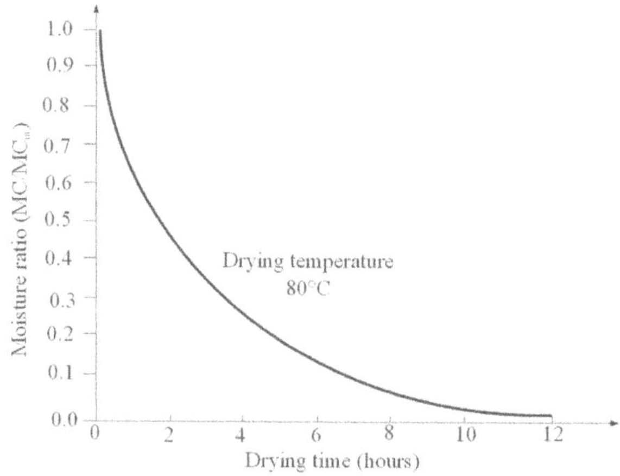

FIGURE 3.8 Variation of moisture ratio versus time for drying of sliced mangoes. (Redrawn from Mercer 2014.).

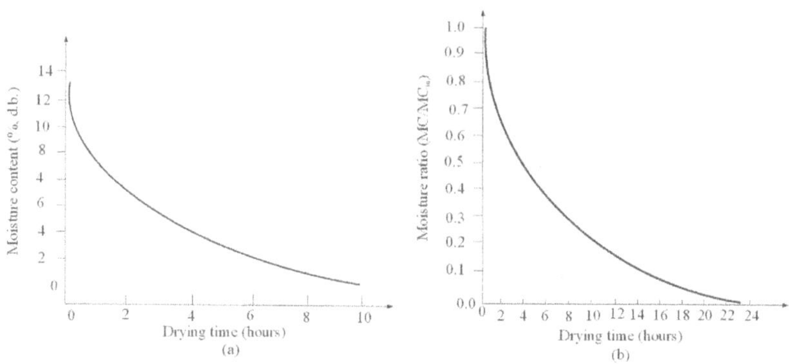

FIGURE 3.9 (a) Variation of moisture content with time of pineapple slices under convective drying. (Redrawn from Rani and Tripathy 2020.) (b) Variation of moisture ratio versus drying time for pineapple slices under convective drying. (Redrawn from Mercer 2014.)

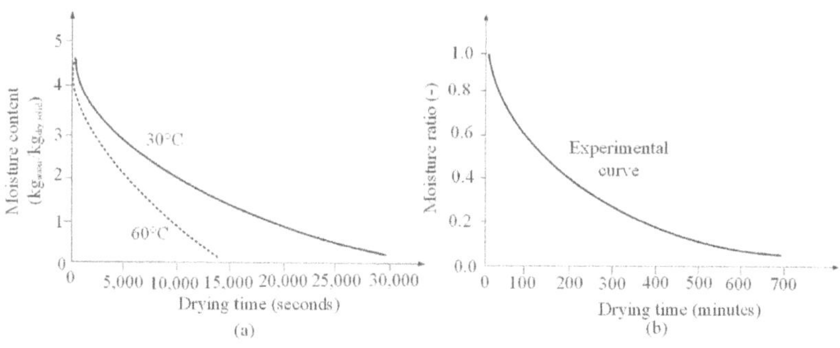

FIGURE 3.10 (a) Variation of moisture content with time of kiwifruit at two different temperatures. (Redrawn from Simal et al. 2005.) (b) Variation of moisture ratio of kiwifruit during hot air drying. (Redrawn from Orikasa et al. 2008.)

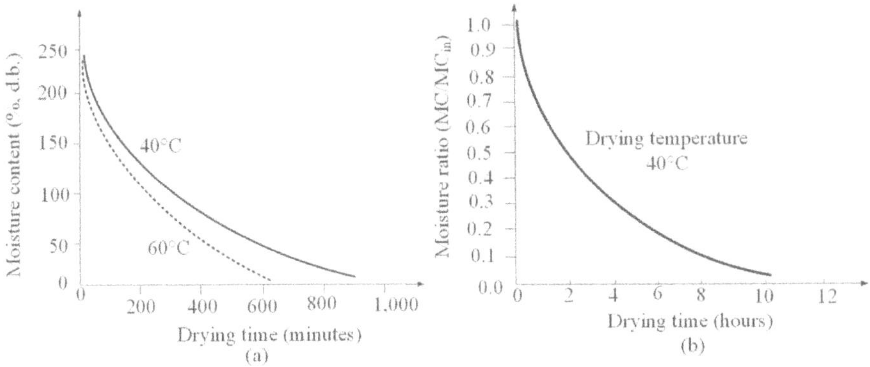

FIGURE 3.11 (a) Variation of moisture content (dry basis) with drying time of stone apple slices dried by convective air drying at 40°C and 60°C. (b) Variation of moisture ratio versus time of royal gala apple rigs. (Redrawn from Mercer 2014.)

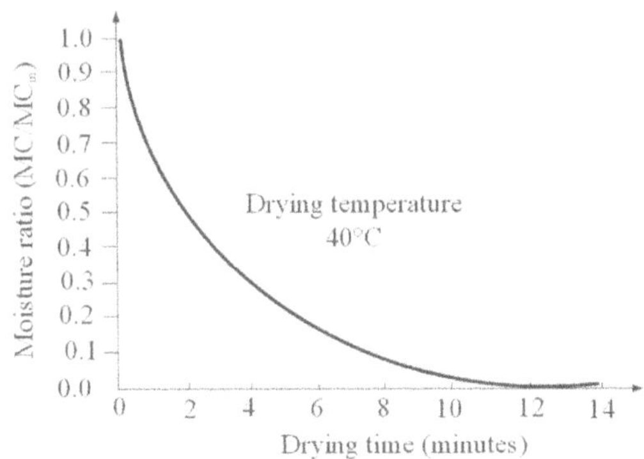

FIGURE 3.12 Variation of moisture content (dry basis) with drying time of carrot slices. (Redrawn from Mercer 2014.)

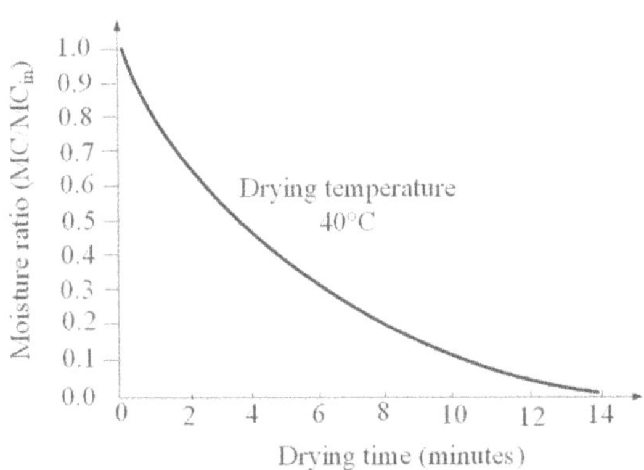

FIGURE 3.13 Variation of moisture ratio with drying time of sweet green pepper. (Redrawn from Mercer 2014.)

TABLE 3.5
Approximate Drying Times of Some Fruits and Vegetables

Dried Product	Drying Temperature (°C)	Drying Time (hours)
Fruits		
Apples slices	50–60	6–10
Apricot slices	50–60	8–10
Peaches (halved or slices)	50–60	12–14
Cherries (whole or halved)	50–60	8–10
Mango slices	50–60	8–12
Banana (whole, halved, coins)	50–60	8–12
Plums	60–70	18–20
Dates (whole)	60–70	18–20
Pineapple rings	60–70	16–17
Papaya slices	60–70	14–16
Pitaya slices	60–70	10–13
Vegetables		
Carrot slices	50–60	10–12
Sweet green pepper slices	50–60	18–20
Tomato wedges	50–60	15–17

3.9 EXPERIMENTAL DRYING TIMES FOR SOME FRUITS AND VEGETABLES

Table 3.5 gives experimental drying times for some fruits and vegetables under conventional hot air convective drying conditions (Mercer 2014).

3.10 TYPICAL DRYING SCHEDULES

Drying schedules are compromises between the need to dry porous solid materials as fast as possible (i.e., achieving satisfactory drying rates) and avoid developing excessive stresses and drying defects (such as checking, cracking, and discoloration).

Drying schedules are usually divided into steps (stages) with different air dry-bulb and wet-bulb temperatures, relative humidity (which is a measure of the amount of water vapor in the air expressed as a percentage of the total amount contained in saturated air at a given temperature and pressure), velocity, and time durations (which mainly depend on the nature and size of dried products). The environment of drying enclosures (which are chosen to achieve satisfactory drying rates and avoid as much as possible drying defects) have to be changed according to the evolution of actual moisture contents in the dried products. In most drying schedules, the air dry-bulb temperatures are gradually increased while the air absolute humidity and wet-bulb temperature progressively decrease.

Traditionally, the drying schedules have been developed based on empirical trial-and-error studies from small-scale tests involving changes of some pertinent parameters. In practice, dryer operators periodically measure the moisture contents of a number of representative material samples.

To develop drying schedules, some of the following basic rules should be considered: (i) type of dryer; (ii) material type; (iii) thickness of material dried pieces (because the drying time is approximately inversely related to thickness); and (iv) intended use of dried material.

3.11 PUBLISHED RESULTS ON FRUITS AND VEGETABLES DRYING

The information provided in this section should serve as starting data to optimize the design of heat pump-assisted drying systems.

3.11.1 Drying of Bananas

Bananas (*Musa*) is an elongated and curved peel fruit that has soft, pulpy flesh. They are low in calories and rich in nutrients such as potassium, vitamin C, vitamin B6, and dietary fiber. In some countries, cooking bananas are called plantains, distinguishing them from dessert bananas.

Bananas also contain water, water vapor, air, and solid porous matrix. During dehydration of unripe and ripe banana samples, the moisture is transferred from their centers to surfaces (Mercer 2014). For drying, bananas are usually peeled and cut into slices of about 5–6 mm thick. To ensure that the drying air contacts all surfaces, the banana slices should be placed on flexible plastic meshes. The value of activation energy (i.e., minimum energy requirement for starting the drying process) for banana slices was found to be 32.65 kJ/mol (Doymaz 2010). In practice, banana slices are dried at temperatures between 50°C and 80°C, with horizontal air velocity of 2.4 m/s by using hot air convective dryers equipped with electrical heaters, adjustable centrifugal fans, and proportional temperature controllers. They are dried up to final moisture contents between $0.012\,kg_{water}/kg_{dry\,solid}$ and $0.020\,kg_{water}/kg_{dry\,solid}$.

Figure 3.14 shows the variation of moisture ratio of banana slices dried at 50°C (Doymaz 2010). The time required to reduce moisture content of banana slices from the initial $60\% \pm 0.5\%$ ($kg_{water}/kg_{dry\,solid}$) (w.b.) to a final $10\% \pm 0.5\%$ $kg_{water}/kg_{dry\,solid}$ (w.b.) are about 180 minutes (at 80°C drying temperature) and 525 minutes (at 50°C drying temperature), respectively.

Empirically, to reach a final moisture content of 10% (wet basis) (i.e., $0.111\,g_{water}/g_{dry\,solids}$) from an initial moisture content of 73% (wet basis) (i.e., $2.80\,g_{water}/g_{dry\,solids}$) of banana slices, Equation 3.4 can be applied (with $k_{banana} = 0.318$):

$$\begin{aligned} t &= \ln\left(MC_{in}/MC\right)/0.318 = \ln\left(2.80/0.111\right)/0.318 \\ &= \ln(25.22)/0.318 = 3.23/0.318 = 10.2\,\text{hours} \end{aligned} \tag{3.8}$$

Therefore, drying the banana slices under these conditions may theoretically take about 10.2 hours.

3.11.2 Drying of Untreated Ripe Green Mango

Mango (*Mangifera indica* L.) is a 7.5–13 cm long (when raped) fruit having significant nutritional and economic importance. It is one of the most important

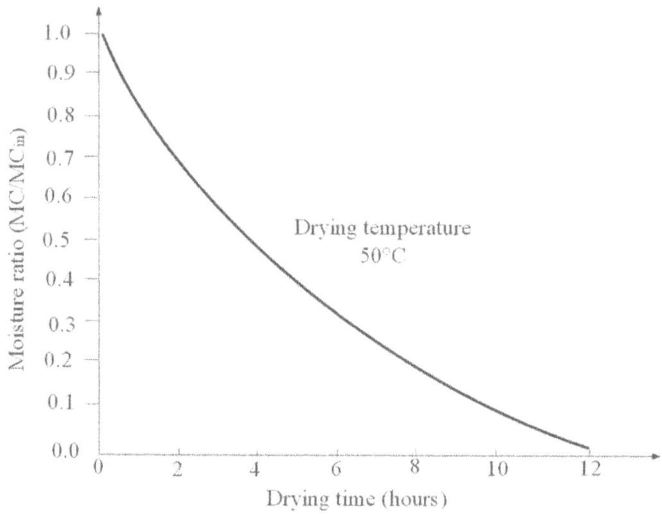

FIGURE 3.14 Variation of moisture ratio of banana slices versus drying time at 50°C. (Redrawn from Doymaz 2010.)

commercialized fruits worldwide in terms of production, marketing, and consumption. It is a plentiful source of vitamin A, B, and C and a rich source of nutrients, minerals, fiber, phytonutrients (as β-carotene), carbohydrates, and calorie. It is low in protein, fat, and dietary fiber, and a potential source of natural antioxidant (Nigam et al. 2007).

Though mangoes are rich in antioxidants and phytonutrients, they have a short shelf life due to high moisture content (80%, w.b.) being highly susceptible to microbial growth and spoilage. Ripe mango without any treatment or processing deteriorates by severe fungal damage within 18 days at 25°C. Moreover, mangoes are seasonal crops, having overproduction during peak season and scarcity during off-season. In some countries (e.g., India), the annual postharvest losses for mangoes are about 12.74% (Sehrawat and Nema 2017).

In spite of its excellent properties, the perishable nature of mangoes and their short harvest season severely limit utilization. Consequently, after harvesting, excessive mango moisture content must be reduced by different drying techniques such as solar, convective hot air, vacuum, freezing, and superheated steam.

Untreated green mangoes are usually peeled, cut into slices (3, 5, or 7 mm thick), placed in stainless steel trays with round perforations in single and double layers, and dried with hot air at 50°C and 60°C. The pieces should have small amounts of spaces between them to ensure that the drying air contacts all surfaces. Air is blown by fans fixed at the bottom of the drying chambers. The velocity of the drying air is normally constant at around 2.25 m/s.

Effective moisture diffusivity and activation energy, estimated using Fick's second law of moisture diffusion and the Arrhenius-type equation, respectively, vary between 1.131×10^{-7} and $1.259\times10^{-7}\,m^2$ (Corzo et al. 2008) and between 43.873 kJ/mol and 44.403 kJ/mol, respectively (Mishra et al. 2012).

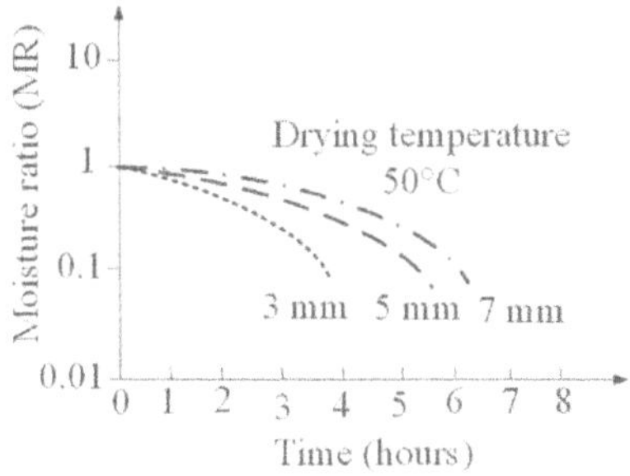

FIGURE 3.15 Average drying curves for untreated green mango slices dried at 50°C. (Redrawn from Mishra et al. 2012.)

The experiments showed that the moisture removal rate increases while increasing the drying temperature and that the drying time decreases as the drying temperature increases. For instance, at 50°C, for untreated mango samples, the moisture content reaches 24.43% (d.b.) after 540 minutes from the initial moisture content of about 507.91% (d.b.). Similarly, at 60°C, the moisture content of untreated mango samples reaches 21.07% (d.b.) after 480 minutes (Akoy 2014; Kabiru et al. 2013).

The moisture ratio of constant thickness mango samples decreases with increasing temperature. Increased drying temperature also give faster drying rates (Karel et al. 1975). To determine the effects of mangos' thickness (e.g., 3, 5, and 7 mm thick) on moisture ratio, experiments were conducted at a temperature of 50°C and constant airflow rates. It can be seen that (Figure 3.15) 3 mm thick mango samples dried at a constant temperature of 50°C requires the least time, while the highest drying time is required for 7 mm thick samples (Mishra et al. 2012). It was also observed that drying time is lower when higher temperature of 65°C is used for a given thickness (Kabiru et al. 2013).

Empirically, to reach a final moisture content of 10% (wet basis) (i.e., $0.111 g_{water}/g_{dry\,solids}$) of untreated green mango slices from an initial moisture content of 85.7% (wet basis) (i.e., $5.97\, g_{water}/g_{dry\,solids}$), Equation 3.4 can be applied (with $k_{mango} = 0.338$):

$$\begin{aligned} t &= \ln\left(MC_{in}/MC\right)/0.338 = \ln\left(5.97/0.111\right)/0.338 \\ &= \ln(53.78)/0.338 = 3.98/0.338 = 11.8 \text{ hours} \end{aligned} \tag{3.9}$$

Therefore, drying mango slices under these conditions could take about 11.8 hours.

Among different drying temperatures (60–80°C) used for drying of mango, the least color change was observed in the mango slices (3 mm) at 80°C as lesser time was required to dry the product. It was found that lightness, yellowness value decreased and redness value increased. At all the temperatures used for drying, browning was not observed, but insignificant differences were reported between the fresh and dried samples (Akoy 2014; Caparino et al. 2012).

Figure 3.16 shows that at a given drying temperature (i.e., 50°C), the mango drying rate decreases continuously with drying time. The drying rate is higher at the beginning of the drying cycle than at the end because the moisture diffusion rate decreases toward the final step of drying.

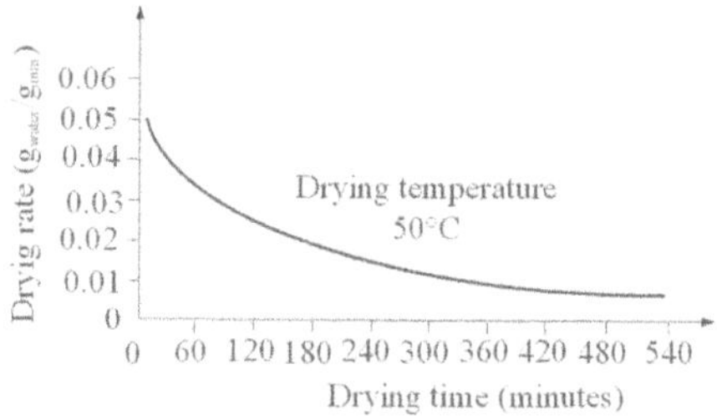

FIGURE 3.16 Average drying rate variation for untreated green mango slices dried at 50°C. (Redrawn from Mishra et al. 2012.)

3.11.3 Drying of Stone Apple Rings

Stone apple (*Aegle marmelos* Correa) is a round or oval-shaped (7–8.5 cm in diameter) fruit with thin red skins surrounding juicy flesh. It has rich nutritional and medicinal properties (Kaya et al. 2007; Rayaguru and Routray 2012).

Dehydration is the best method for the preservation of stone apple fruit pulp, because it decreases the moisture content. Thin layer drying characteristics of stone apple slices are usually investigated in hot air convective dryers consisting of centrifugal blowers, electrical resistances, measurement sensors, and data recording devices. The blowers and heaters of dryers are switched on for 30 minutes for the drying air to reach stable temperatures.

Drying characteristics of stone apple slices dried in hot air convective dryers show that to reduce the initial moisture content of 160–235% (d.b.) to final moisture content of 6–7% (d.b.), the total times required for drying at 40°C and 60°C are 840 minutes and 540 minutes, respectively (Rayaguru and Routray 2012).

For drying, stone apples are sliced to 8 mm thickness, circular slices of diameter 10–15 mm, and then dried at temperatures varying from 40°C to 60°C in forced hot air convection dryers. Experiments are usually carried out at maximum 1.1 ± 0.2 m/s air velocity.

Activation energy of stone apple slices (calculated using an exponential expression based on the Arrhenius equation) (Figure 3.17) was found to be 16.1 kJ/mol. Effective moisture diffusivity values range from $3.7317 \times 10^{-10}\,m^2/s$ at 40°C to $6.675 \times 10^{-10}\,m^2/s$ at 60°C.

Figure 3.18 shows that at higher drying temperatures (60°C versus 40°C), the drying curves exhibit steeper slope, and thus higher drying rates.

Figure 3.19 shows the evolution of moisture contents of stone apple slices dried at 40°Cand 60°C as functions of drying time.

Empirically, to reach a final moisture content of 10% (wet basis) (i.e., $0.111\,g_{water}/g_{dry\,solids}$) of stone apple slices from an initial moisture content of 85.4% (wet basis) (i.e., $5.84\ g_{water}/g_{dry\,solids}$), Equation 3.4 can be applied (with $k_{apple} = 0.396$):

$$t = \ln\left(MC_{in}/MC\right)/0.396 = \ln\left(5.84/0.111\right)/0.396$$
$$= \ln(52.61)/0.396 = 3.96/0.396 = 10.0\,\text{hours} \tag{3.10}$$

Therefore, drying stone apple rings under these conditions should take about 10 hours.

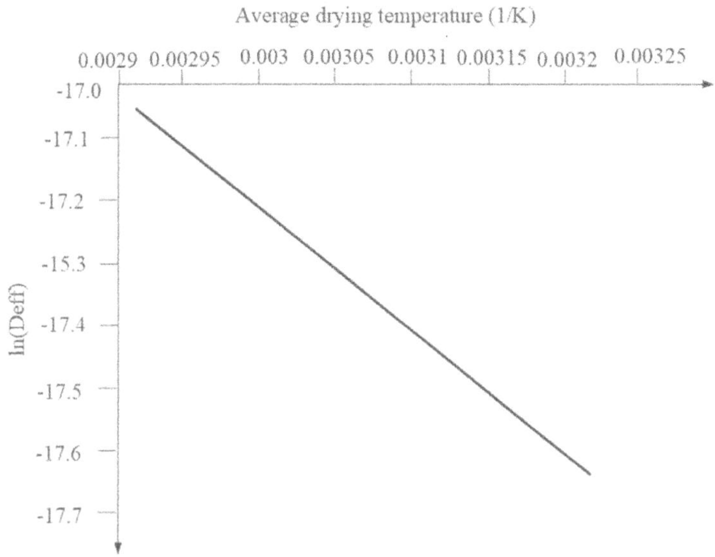

FIGURE 3.17 Effective moisture diffusivity plotted as a function of drying temperature used for estimation of activation energy (Rayaguru and Routray 2012).

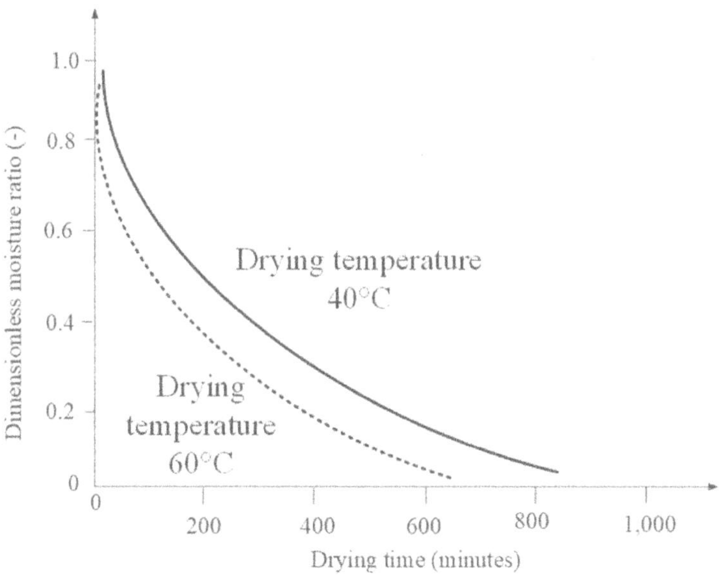

FIGURE 3.18 Variation of moisture content (dry basis) with drying time of stone apple slices dried by convective air drying at 40°C and 60°C. (Redrawn from Rayaguru and Routray 2012.)

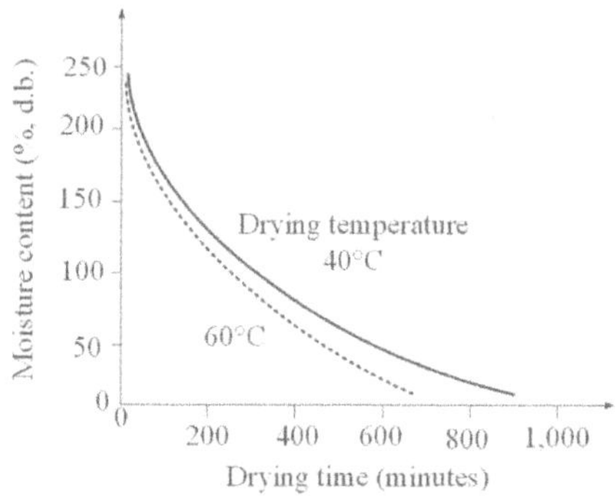

FIGURE 3.19 Variation of moisture contents with drying time of stone apple slices dried at 40°C and 60°C.

3.11.4 Drying of Figs

Figs (*Ficus carica*) are 3–5 cm long purple fruits that have soft and chewy textures with slight crunches from the seeds.

Experiments intending to determine the drying characteristics of figs have been conducted with drying air at temperatures ranging from 55°C to 85°C and the air velocities in the range of 0.5–3 m/s (Babalis and Belessiotis 2004). The effective diffusivity (D_{eff}) of dried figs vary from 8.40×10^{-10} to 1.13×10^{-9} m^2/s depending on different values of drying air temperatures and air velocities. Maximum value of the effective diffusivity has been achieved at an air velocity of 1.5 m/s. The calculated energy of activation varied from a minimum of 30.8 kJ/mol to a maximum of 48.47 kJ/mol exhibiting an apparent minimum value at approximately 1.5 m/s.

Figure 3.20 shows the influence of drying time and drying air temperature on figs moisture ratios when the air velocity is kept constant at 2 m/s.

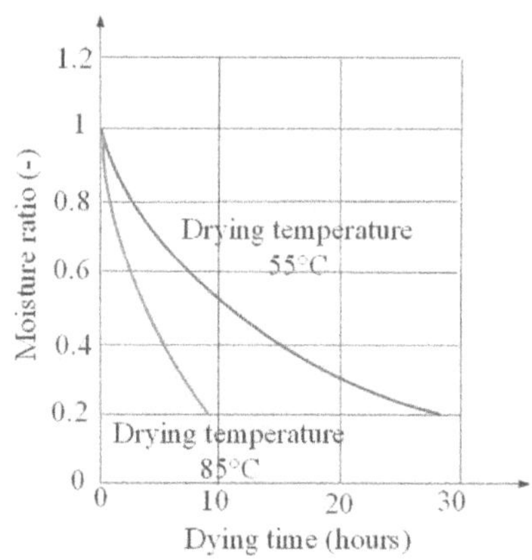

FIGURE 3.20 Effect of drying air temperatures on moisture ratios of dried figs for drying air velocity of 2 m/s. (Redrawn from Babalis and Belessiotis 2004.)

It can be seen that the values of moisture ratios decrease rapidly, with consequent increase of the drying rate, when the drying air temperature increase. The initial values of drying rates almost double when the drying air temperature increases from 55°C to 85°C.

At a constant drying temperature of 85°C, the influence of air velocity is significant at the beginning of drying. At an air velocity of 3 m/s, the drying rate is about 150 g_{water} / h, almost three times the corresponding value for 0.5 m/s, implying that the evaporation process took place at the surface and, therefore, is more directly affected by the drying air velocity. Subsequently, and after a short period of time (4–5 hours), the predominance of the air velocity variation on the drying rate become less important.

In conclusion, by studying the role of drying air temperature in the range of 55–85°C and of the air velocity in the range of 0.5–3 m/s, it was found that a range of 1–2 m/s for the drying air velocity could be optimum values inside industrial dryers, providing the best cost/efficiency performance ratio for the drying of figs.

3.12 PUBLISHED RESULTS ON DRYING VEGETABLES

Many theoretical and experimental works were conducted to understand and describe the drying process for different vegetable products containing between 80% and 90% moisture (water). The results have been used to estimate drying times under different drying conditions, how to increase the drying efficiency, and, also to generalize the respective processes (Ertekin and Ziya Firat 2017).

3.12.1 Drying of Carrots

Carrots (*Daucus carota*) are crunchy, tasty, and highly nutritious root vegetables. They are good sources of β-carotene, fiber, vitamin K, potassium, and antioxidants, while containing low calories and low cholesterol levels (Mercer 2014).

Carrot slices are usually dried at a temperature of about 500°C and constant air velocity of 1 m/s. The pieces to dry should have small amounts of space between them to ensure the drying air contacts all surfaces. Plastic meshes can be used to prevent the carrot slices from falling through the wire mesh rack as they dry (Aghbashlo et al. 2009).

Figure 3.21 shows a typical moisture ratio variation of dried sliced carrots.

Empirically, to reach a final moisture content of 10% (wet basis) (i.e., 0.111 $g_{water}/g_{dry\,solids}$) of sliced carrots from an initial moisture content of 89% (wet basis) (i.e., 8.10 $g_{water}/g_{dry\,solids}$), Equation 3.4 can be applied (with $k_{carrots} = 0.409$):

$$
\begin{aligned}
t &= \ln\left(MC_{in}/MC\right)/0.409 = \ln\left(8.10/0.111\right)/0.409 \\
&= \ln(72.97)/0.409 = 4.29/0.409 = 10.5\,\text{hours}
\end{aligned}
\tag{3.11}
$$

Therefore, drying carrot slices under these conditions could take about 10.5 hours.

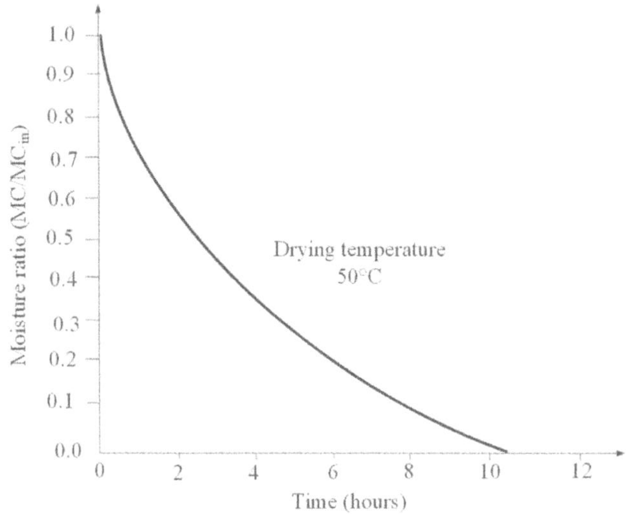

FIGURE 3.21 Typical drying curve of sliced carrots. (Redrawn from Aghbashlo et al. 2009.)

3.12.2 Drying of Ginger

Ginger root (*Zingiber officinale*) is a flowering plant widely used as a spice and a medical herb (https://en.wikipedia.org/, accessed August 23, 2024).

Optimal drying conditions of sliced ginger roots are experimentally determined in laboratory hot air convective dryers at different temperatures (e.g., 400°C, 500°C, 600°C, and 700°C) and thickness (e.g., 2, 5, 10, and 15 mm), velocities (e.g., 0.8, 1.5, and 3 m/s), and loading densities (e.g., 5 kg/m^2).

It was found that (Afolabi et al. 2014; Akinola et al. 2018; Akpinar and Toraman 2015; Bai et al. 2023; Shaukat et al. 2024) (i) increasing the drying air temperature causes shorter drying times; (ii) thinner ginger slices causes shorter drying times; (iii) effective moisture diffusivity values increase from 2.807×10^{-10} m^2/s to 6.977×10^{-9} m^2/s; (iv) activation energy values for different slice thickness of ginger vary from 19.15 kJ/mol to 22.79 kJ/mol; (v) total needed energy varies from 735.3 kWh to 868.5 kWh; (vi) specific energy requirements vary from 3676.6 kWh/kg to 4342.4 kWh/kg; and (vii) convective heat transfer coefficients change from 0.33 W/m^{2}K to 2.11 W/m^{2}K.

Figure 3.22 shows a typical variation of moisture ratios of sliced ginger roots dried at 50°C.

Empirically, to reach a final moisture content of 10% (wet basis) (i.e., $0.111 g_{water}/g_{dry\,solids}$) of dried sliced ginger roots from an initial moisture content of 89.7% (wet basis) (i.e., $8.71 g_{water}/g_{dry\,solids}$), Equation 3.4 can be applied (with $k_{ginger} = 0.601$):

$$
\begin{aligned}
t &= \ln\left(MC_{in}/MC\right)/0.601 = \ln(8.71/0.111)/0.601 \\
&= \ln(77.5)/0.601 = 4.36/0.601 = 7.27\,\text{hours}
\end{aligned} \tag{3.12}
$$

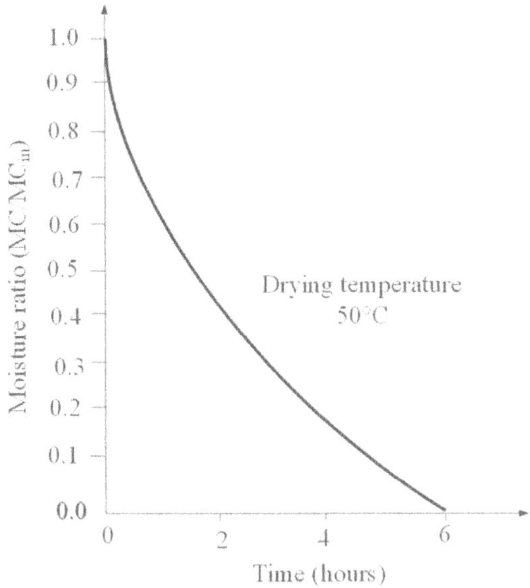

FIGURE 3.22 Typical variation of moisture ratios of sliced ginger roots. (Redrawn from Bai et al. 2023.)

Therefore, drying sliced ginger roots under these conditions could take about 7.1 hours.

3.12.3 Sweet Green Pepper

The term "sweet pepper" refers to a range of pepper varieties with mild heat and fruity or sweet flavor, including bell peppers, banana peppers, and pimentos.

Sweet green pepper slices are placed on a laboratory dryer rack with the skin side down (i.e., touching the rack) and the fleshy portion pointed upward to increase the exposure of the moist, porous inner surface to drying air and improve the overall efficiency of the drying process.

Experiments with drying air at a temperature of about 500°C and linear velocity of 0.5 m/s have been performed (Mercer 2014). Figure 3.23 represents a typical drying curve of dried sweet green pepper slices.

Empirically, to reach a final moisture content of 10% (wet basis) (i.e., $0.111 g_{water}/g_{dry\,solids}$) of sweet green pepper slices dried from an initial moisture content of 94.4% (wet basis) (i.e., $16.74\, g_{water}/g_{dry\,solids}$), Equation 3.4 can be applied (with $k_{green\,papper} = 0.251$):

$$\begin{aligned} t &= \ln\left(MC_{in}/MC\right)/0.251 = \ln\left(16.74/0.111\right)/0.251 \\ &= \ln(150.8)/0.251 = 5.02/0.252 = 20.0\,\text{hours} \end{aligned} \tag{3.13}$$

Therefore, drying sweet green pepper slices under these conditions could take about 20 hours.

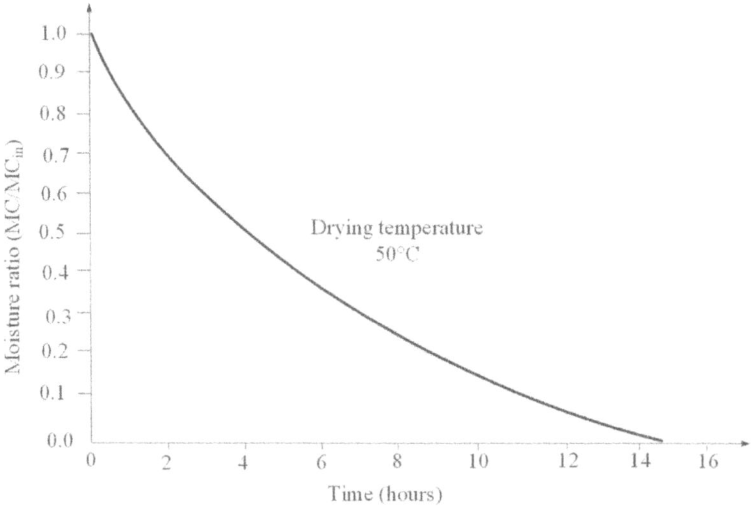

FIGURE 3.23 Typical variation of moisture content of dried sweet green pepper slices. (Redrawn from Mercer 2014.)

3.12.4 Tomato Wedges

Tomatoes (*Lycopersicon esculentum*) are a good source of vitamins C and K, potassium, and folate.

Tomato wedges are usually placed on dryer racks with the skin sides down and small amounts of spaces between them to ensure the drying air contacts all surfaces.

Nine different mathematical models of tomatoes dried with air at a temperature of about 500°C, linear air velocity of 0.5 m/s, and tomato wedge loads of 2–6 kg/m^2 showed that dryer efficiency increases from 17.33% to 35% with increase in drying loads from 2 kg/m^2 to 6 kg/m^2 (Arepally et al. 2017):

Figure 3.24 shows a typical variation of moisture ratio (drying curve) with time of tomato wedges dried at 50°C (Arepally et al. 2017).

Other experiments have been performed with 1 kg batch of freshly harvested tomato samples sliced to 10 mm thickness and dried at varying air velocities (1.0, 1.5, and 2.0 m/s) and temperatures (50°C, 55°C, 60°C, 65°C, and 70°C) (Nwakuba et al. 2018).

The results showed that (i) the effective moisture diffusivity increased with increase in temperature and air velocity; the highest value was obtained at air velocity of 2.0 m/s and temperature of 70°C, whereas it was lowest at air velocity of 1.0 m/s and temperature of 50°C; (ii) the mean activation energy required to dry 1 kg of 10 mm sliced tomato samples was 39.34 kJ/mol; (iii) energy consumption varied between 5.42 kWh and 99.78 kWh; (iv) the specific energy consumption varied between 5.53 kWh/kg and 150.61 kWh/kg; (v) at optimum drying conditions of

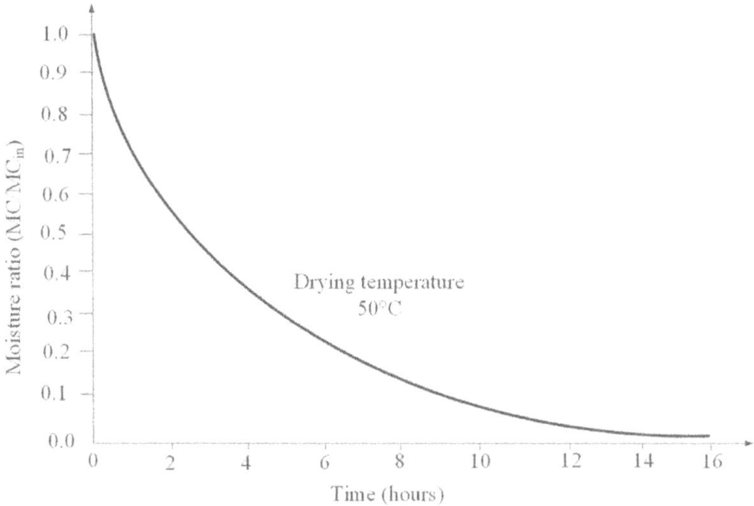

FIGURE 3.24 Typical drying curve of tomato wedges. (Redrawn from Arepally et al. 2017.)

1.94 m/s air velocity, 10.36 mm slice thickness, and 68.4°C drying air temperature, the corresponding energy consumption was 5.68 kWh.

Empirically, to reach a final moisture content of 10% (wet basis) (i.e., $0.111 g_{water}/g_{dry\ solids}$) from an initial moisture content of 93.8% (wet basis) (i.e., $15.08\ g_{water}/g_{dry\ solids}$), Equation 3.4 can be applied (with $k_{tomato} = 0.287$):

$$\begin{aligned} t &= \ln\left(MC_{in}/MC\right)/0.287 = \ln\left(15.08/0.111\right)/0.287 \\ &= \ln(135.9)/0.287 = 4.91/0.287 = 17.1 \text{ hours} \end{aligned} \quad (3.14)$$

Therefore, drying tomato wedges under these conditions should take about 17.1 hours.

3.12.5 Drying of Potato Slices

Potato (*Solanum tuberosum*) is the fourth most important vegetable crop for human nutrition in the world that contains vitamin C, protein, thiamin, and niacin (https://www.britannica.com/, accessed March 4, 2025). It is also an important source of carbohydrate, β-carotene, fiber, and potassium ions.

Experiments of drying of potato slices were conducted at air temperatures between 45°C and 70°C with air velocities of 1.60 m/s and 1.81 m/s.

Figure 3.25 represents the variation of moisture ratios of circular potato samples at drying temperature of 55°C and 1.60 m/s air velocity. It can be seen that the moisture ratio reaches the value of 0.11 after 256 minutes.

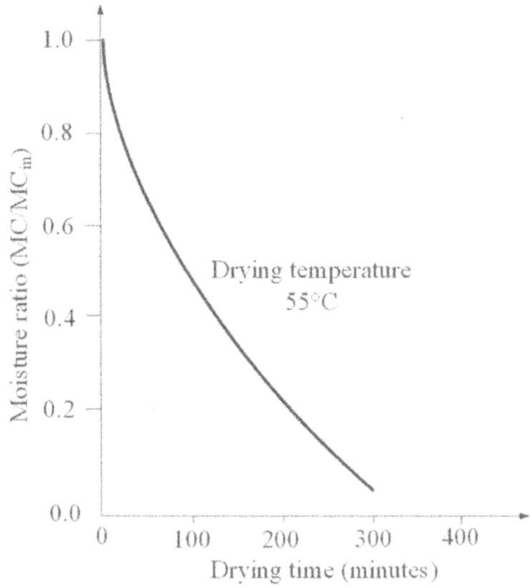

FIGURE 3.25 Typical variation of moisture ratio of potato slices dried at 55°C.

REFERENCES

Afolabi, T., T. Tunde-Akintunde, O. Oyelade. 2014. Influence of drying conditions on the effective moisture diffusivity and energy requirements of ginger slices. *Journal of Field Robotics* (https://www.semanticscholar.org/, accessed June 18, 2024).

Aghbashlo, M., M.H., Kianmehr, S. Khani, M. Ghasemi. 2009. Mathematical modelling of thin layer drying of carrot. *International Agrophysics* 23:313–317.

Akinola, A., O. Azeta, S.N. Ezeorah. 2018. Evaluation of the dehydration characteristics of ginger (*Zingiber officinale*) root slices using refractance window drying technology. *Agricultural and Food Sciences* 3(1) (https://www.mendeley.com/, accessed August 25, 2024).

Akoy, E.O.M. 2014. Effect of drying temperature on some quality attributes of mango slices. *International Journal of Innovation and Scientific Research* 4(2):91–99.

Akpinar, E., S. Toraman. 2015. Determination of drying kinetics and convective heat transfer coefficients of ginger slices. *Agricultural and Food Sciences* 52(2):271–281.

Arepally, D., S.R. Ravula, V.R. Kamidi. 2017. Mathematical modelling of mixed mode natural convection solar drying of tomato slices. *International Journal of Chemical Studies* 5(4):1274–1279.

Babalis, S.J., V.G. Belessiotis. 2004. Influence of the drying conditions on the drying constants and moisture diffusivity during the thin-layer drying of figs. *Journal of Food Engineering* 65(3):449–458.

Bai, J.W., Z.J. Gao, H.W. Xiao, X.T. Wang, Q. Zhang. 2013. Polyphenol oxidase inactivation and vitamin C degradation kinetics of Fuji apple quarters by high humidity air impingement blanching. *International Journal of Food Science and Technology* 48(6):1135–1141.

Bai, R., J. Sun, B. Zhang. 2023. Hot air convective drying of ginger slices: Drying behavior, quality characteristics, optimization of parameters, and volatile fingerprints analysis. *Food* 12(6):1283.

Berk, Z. 2009. Freeze drying (lyophilisation) and freeze concentration. In: *Food Process Engineering and Technology*, edited by Z. Berk, Academic Press, San Diego, CA, pp. 511–523.

Biesaga, M. 2011. Influence of extraction methods on stability of flavonoids. *Journal of Chromatography* 1218 (18):2505–2512.

Caparino, O.A., J. Tang, C.I. Nindo, S.S. Sablani, J.R. Powers, J.K. Fellman. 2012. Effect of drying methods on the physical properties and microstructures of mango (*Philippine 'Carabao' var.*) powder. *Journal of Food Engineering* 111:135–148.

Chan, E.W.C., Y.Y. Lim, S.K. Wong, K.K. Lim, S.P. Tan, F.S. Lianto, M.Y. Yong. 2009. Effects of different drying methods on the antioxidant properties of leaves and tea of ginger species. *Food Chemistry* 113(1):166–172.

Corzo, O., N. Bracho, C. Alvarez. 2008. Water effective diffusion coefficient of mango slices at different maturity stages during air drying. *Journal of Food Engineering* 87:479–484.

Doymaz, I. 2010. Evaluation of mathematical models for prediction of thin-layer drying of banana slices. *International Journal of Food Properties* 13:486–497.

Ertekin, C., M. Ziya Firat. 2017. A comprehensive review of thin-layer drying models used in agricultural products. *Critical Reviews in Food Science and Nutrients* 57(4):701–717.

Kabiru, A.A., A.A. Joshua, A.O. Raji. 2013. Effect of slice thickness and temperature on the drying kinetics of mango (*Mangifera indica*). *International Journal of Recent Research and Applied Studies* 15(1):41–50.

Karel, M., O.R. Fennema, D.B. Lund. 1975. *Principles of Food Science: Part-II—Physical Principles of Food Preservation*, Marcel Dekker, Inc., New York, NY.

Kaya, A., O. Aydin, C. Demirtas. 2007. Drying kinetics of red delicious apple. *Biosystems Engineering* 96:517–524.

Lewicki, P.P. 2006. Design of hot air drying for better foods. *Trends in Food Science and Technology* 17(4):153–163.

Maskan, M. 2001. Drying, shrinkage and rehydration characteristics of kiwifruits during hot air and microwave drying. *Journal of Food Engineering* 48(2):177–182.

Mercer, D.G. 2014. *An Introduction to the Dehydration and Drying of Fruits and Vegetables: Drying of Specific Fruits and Vegetables*: Part 2. Department of Food Science, University of Guelph, Ontario, Canada.

Miller, N.J., C.A. Rice-Evans. 1997. The relative contributions of ascorbic acid and phenolic antioxidants to the total antioxidant activity of orange and apple fruit juices and black-currant drink. *Food Chemistry* 60(3):331–337.

Mishra, M., Kandasamy, P., Shukla, R.N., Kumar, A. 2012. Convective hot-air drying of green mango: Influence of hot water blanching and chemical pretreatment on drying kinetics and physicochemical properties of dried product. *International Journal of Fruit Science* 21(1):732–757.

Mitra, J., S.L. Shrivastava, P. Srinivasa Rao. 2011. Process optimization of vacuum drying of onion slices. *Czech Journal of Food Science* 29(6):586–594.

Mujumdar, A.S. 2014. *Handbook of Industrial Drying*, 4th edition, CRC Press, Boca Raton, FL.

Mujumdar, A.S., S. Devahastin. 2000. *Fundamental principles of drying*. In: *Mujumdar's Practical Guide to Industrial Drying*, edited by S. Devahastin, Exergex Corporation, Brossard, Quebec.

Nigam, S., D.K. Bhatt, A. Jha. 2007. Different product of mango: The king of fruits. *Processed Food Industry* 10(9):32–40.

Nunes, J.C., M.G. Lago, V.N. Castelo-Branco, F.R. Oliveira, A.G. Torres, D. Perrone, M. Monteiro. 2016. Effect of drying method on volatile compounds, phenolic profile and antioxidant capacity of guava powders. *Food Chemistry* 197:881–890.

Nwakuba, N., O. Chukwuezie, S.N. Asoegwu, G. Nwandikom, N. Okereke. 2018. Thin layer modeling and determination of thermodynamic properties of tomato slices during hot air drying. *Poljoprivredna Tehnika* 43(1):39–51.

Okos, M.R., G. Narsimhan, R.K. Singh, A.C. Weitnauer. 1992. *Food dehydration*. In *Handbook of Food Engineering*, edited by D.R. Heldman and D.B. Lund, Marcel Dekker, New York, NY.

Orikasa, T., L. Wu, T. Shiina, A. Tagawa. 2008. Drying characteristics of kiwifruit during hot air drying. *Journal of Food Engineering* 85:303–308.

Rani, P, Tripathy, P.P. 2020. Modelling of moisture migration during convective drying of pineapple slice considering non-isotropic shrinkage and variable transport properties. *Journal of Food Science and Technology* 57:3748–3761.

Rayaguru, K., W. Routray. 2012. Mathematical modeling of thin layer drying kinetics of stone apple slices. *International Food Research Journal* 19(4):1503–1510.

Reis, F.R. (Ed.). 2014. *Vacuum Drying for Extending Food Shelf-Life*, Springer Briefs in Applied Sciences and Technology, Springer, Cham, Switzerland.

Rodríguez, R., J.I. Lombraña, M. Kamel, C. de Elvira. 2005. Kinetic and quality study of mushroom drying under microwave and vacuum. *Drying Technology* 23(9–11):2197–2213.

Sagar, V.R., P.S. Kumar. 2010. Recent advances in drying and dehydration of fruits and vegetables: A review. *Journal of Food Science and Technology* 47(1):15–26.

Sehrawat, R., P.K. Nema. 2017. Drying of Mango Products. In: *Processing and Drying of Tropical Fruits*, 1st edition, edited by C.L., Law, C.L. Hii, S.V. Jangam, A.S. Mujumdar, pp. 123–144. Singapore.

Shaukat, M.N., B. Fallico, A. Nazir. 2024. Impact of air-drying temperatures on drying kinetics, physicochemical properties, and bioactive profile of ginger. *Foods* 13:1096–1112.

Simal, S., A. Femenia, M.C. Garau, C. Rosselló. 2005. Use of exponential, Page's and diffusion models to simulate the drying kinetics of kiwi fruit. *Journal of Food Engineering* 66:323–328.

Tiris, C., N. Ozbalta, Tiris, M., I. Dincer. 1994. Performance of a solar dryer. *Energy* 19:993–997.

Vinson, J.A., X. Su, L. Zubik, P. Bose. 2001. Phenol antioxidant quantity and quality in foods: Fruits. *Journal of Agricultural and Food Chemistry* 49:5315–5321.

4 Drying of Woods

4.1 INTRODUCTION

In 2020, about half (54%) of the world's forest was found in countries such as the Russian Federation (20.1%), Brazil (12.2%), Canada (8.6%), the United States (7.6%), and China (5.4%) (https://en.wikipedia.org, accessed January 24, 2025).

In the United States and Canada, timber refers to the wood contents of live trees that have been cut down but not yet processed, while lumber typically refers to wood that has been processed as a building material (boards, planks). In the United Kingdom, Australia, and New Zealand, timber is a term used for sawn wood products, meaning the same things as lumber (https://www.dictionary.com/e/timber-vs-lumber, accessed August 28, 2024).

The main objective of industrial wood drying is to add value to the dried lumber by minimizing the development of defects and achieve low and uniform moisture contents suitable for the intended end use and/or before machining, gluing, and finishing are carried out.

Industrial wood drying is the most time- and energy-demanding process at sawmills' production, and an essential step in the manufacturing of most commercial wood products. It has a great impact on parameters that affect the selling price of the sawn timber. In Canada, for example, the wood industry accounts for 3% of the country's gross domestic product and 10% of total foreign exchange. Canadian sawmills usually deliver more than 75 million m^3 of softwood (as white and black spruce, jack pine, and balsam fir), majority of this production being dried with hot air convective batch dryers using bark-fueled (85%) and fossil fuel-burned (natural gas, propane, or oil) boilers (10%).

In the past few decades, industrial wood drying has made great progress through increased knowledge and awareness of both sawmill managers and kiln operators as to how to operate the kilns to achieve the best drying results. The research and development performed by sawmills, kiln manufacturers, and universities has played a major role in that progress. As a result, the kiln design, process control, and drying schedules have considerably improved.

Today, the main objective of the development of drying technology is to further improve the drying processes by the application of advanced drying technologies (as heat pump-assisted drying), as well as new measurement, modeling, and simulation techniques.

4.2 GREEN WOOD TYPES, BASIC CROSS SECTION, AND PRIMARY PROCESSING

Wood, a porous, hygroscopic substance consisting of dead and living cells, and filled with either water or air, can be divided into two classes (Keey et al. 2000): (i) softwood (such as spruce, pine, and fir) and (ii) hardwood (such as oak and maple).

DOI: 10.1201/9781003618386-4

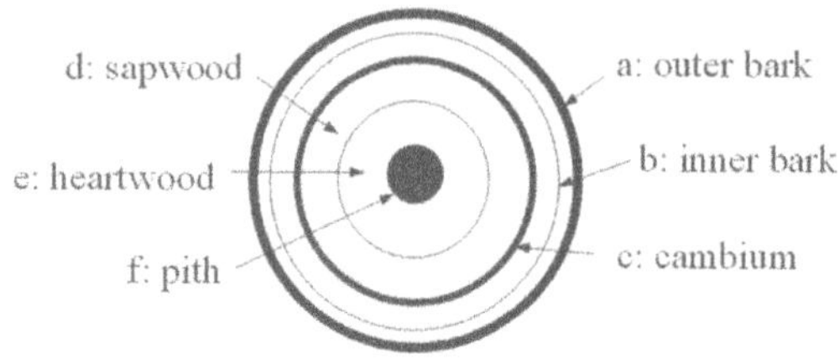

FIGURE 4.1 Schematic cross section of a tree trunk.

According to their ease of drying and drying degradation, timber species can be classified (Cech and Pfaff 2000) as follows: (i) hard-to-dry (hardwoods), (ii) moderately hard-to-dry (mostly exotic hardwoods), and (iii) easy-to-dry (softwoods).

There are three types of cells in woods (Bian 2001): (i) vessels, which are pipe-like structures of indeterminate lengths formed by individual cells connected endwise; (ii) fibers (present only in hardwoods); and (iii) parenchyma containing storage tissues.

Wood cells, which hold the wood structure and transport water and nutrients, are built of three polymers: (i) cellulose (as a structural component accounting for 40–50%), (ii) hemicellulose (which surrounds cellulose microfibers and together forms the wood fiber framework), and (iii) lignin (which binds the cellulose structure together and provides the rigidity and plastic nature of the wood).

From the outermost to the center, a cross section of mature trees consists of (Figure 4.1) (Dinwoodie 1981) (i) outer dead bark (a), (ii) inner bark (b), (iii) cambium (c), (iv) sapwood (d) heartwood (e), and (vi) pith (f).

Sawmills produce lumber from forested green timber with moisture content between 35% and 180% (d.b.) through the following processes (Anderson 2014): (i) when the timber arrives at sawmill, it is sorted and stored in the lumberyard; (ii) bark is separated from the timber before the sawing process; (iii) timber logs are square edged, quarter or radially sawn; (iv) sawn lumber boards are sorted depending on quality and length, pre-dried by natural (yard) or accelerated methods, and, finally, dried in facilities called kilns; and (iv) after drying, the lumber boards are sorted once more and packaged for transportation.

Softwood and hardwood lumber boards made even and cut to standardized width and thickness specified in millimeters are termed as dimensional lumber. Such dimensional green boards are placed inside batch dryers to be dried.

4.3 DRYING PROPERTIES OF WOODS

The properties of living trees are related to directions such as (i) longitudinal (axial) (parallel to the stem that is a hard, rigid structure due to the presence of lignin and secondary growth), (ii) radial (horizontal rays that connect the tree's layers from pith to bark for storage and transfer of nutrients), and (iii) tangential to the bark.

Drying properties of wood include (Glass and Zelinka 2010) (i) density, (ii) specific heat, (iii) thermal conductivity, (iv) moisture content, (v) fiber saturation point, and (vi) equilibrium moisture content (EMC).

4.3.1 Density

The wood's density is a variable parameter influenced by factors such as (i) amount of lignin in cells, (ii) how densely the cells grow together, (iii) number of air spaces within the cells, (iv) amount of moisture content, and (v) chemical composition.

Wood density is defined as the mass of wood per unit of volume of green, oven-dry, or the moisture content in between (Eekelman 1968):

$$\rho_{wood} = m_{wood} / V_{wood} \tag{4.1}$$

where

ρ_{wood} is the wood mass density (kg/m^3);

m_{wood} is the green (kg_{green}), oven-dry ($kg_{oven\,dry}$), or partially dried ($kg_{MC\%}$) mass of wood; and

V_{wood} is the volume of green (m^3_{green}), oven-dry ($m^3_{oven\,dry}$), or partially dried $\left(m^3_{MC\%}\right)$ mass of wood.

Mass density of softwood ranges from $350\,kg/m^3$ to $700\,kg/m^3$, while that of hardwood varies between $450\,kg/m^3$ and $1250\,kg/m^3$.

Wood basic density is defined as the oven-dry mass of a wood sample divided by its green, swollen (saturated) volume (i.e., the green volume of the wood sample when it is in equilibrium with surrounding medium, before any shrinkage occurs due to drying):

$$\rho^{basic}_{wood} = m_{oven-dry} / V_{green} \tag{4.2}$$

where

ρ^{basic}_{wood} is the basic density of wood sample $\left(kg_{oven-dry} / m^3_{green}\right)$;

$m_{oven-dry}$ is the oven-dry mass of wood sample ($kg_{oven-dry}$); and

V_{green} is the wood swollen (saturated) volume of wood sample (m^3_{green}).

Table 4.1 shows the basic densities for some softwood and hardwood species utilized in eastern Canada (Cech and Pfaff 2000).

TABLE 4.1
Basic Density of Some Species of Wood Utilized in Eastern Canada

Species	Basic Density ($kg_{oven\,dry} / m^3_{green}$)
Softwood	
Balsam fir	0.34
White spruce	0.35
Red pine	0.39
Hardwood	
White birch	0.51
Sugar maple	0.60
White oak	0.65

4.3.2 Specific Heat

Specific heat is defined as the amount of heat needed to increase the temperature of 1 kg of wood by 1 K. It depends on the temperature and moisture content of the wood sample but is practically independent of density or species (Glass and Zelinka 2010).

Specific heat of a wood specimen can be determined as follows (Dunlap 1912):

$$\bar{c}_p = 0.266 + 0.00116 \cdot \bar{T} \tag{4.3}$$

where

$\bar{c}_p$ is the average specific heat at constant pressure of wood specimen (kcal/kg) (1 kcal = 4.1868 kJ); and

$\bar{T}$ is the average temperature of the wood specimen over the interval from 0°C to $T_1 = 100$°C.

For a wood sample undergoing a temperature change from T_1 to T_2, where $T_0 \neq 0$°C, the mean specific heat (kcal/kg·K) (1 kcal = 4.1868 kJ) in this range is given by (Beall 1968)

$$\bar{c}_p = 0.266 + 0.00058 \cdot \left[(T_1 + T_2)/2 \right] \tag{4.4}$$

Table 4.2 shows the average specific heats for some wood species.

4.3.3 Thermal Conductivity

Thermal conductivity of wood is a measure of the heat transfer through a sample subjected to unit temperature difference (1 K) across unit thickness (1 m).

Thermal conductivity of wood is affected by factors such as (i) density, (ii) temperature, (iii) moisture content, for example, the thermal conductivity of softwood lumber at 12% (d.b.) moisture content varies in the range of 0.10–0.14 W/m·K, and (iv) structural irregularities, such as checks and knots.

Thermal conductivity increases as density, moisture content, temperature, or extractive content of the wood increases, and is nearly the same in the radial and tangential directions.

TABLE 4.2
Average Specific Heat of Some Wood Species

Species	Specific Heat (J/kg·K)
Birch	1,900
Maple	1,600
Oak	2,400
Red pine	1,500

For moisture content below 25% (d.b.), approximate thermal conductivity across the grain can be calculated with the following linear equation (Glass and Zelinka 2010):

$$k = \rho_{wood}^{basic} \cdot (B + C \cdot MC) + A \tag{4.5}$$

where
k is the thermal conductivity $(\mathrm{W/m \cdot K})$;
ρ_{wood}^{basic} is the basic density of wood sample at moisture content MC $\left(\mathrm{kg}_{oven-dry}/\mathrm{m}^3_{green}\right)$;
MC is the moisture content $(\mathrm{kg}_{water}/\mathrm{kg}_{dry\,matter})$;
A, B, and C are experimental constants; and
for $\rho_{wood}^{basic} \geq 0.3$, temperatures around 24°C, and $MC < 25\%$ (d.b.) the values of the constants are as follows: $A = 0.01864$; $B = 0.1941$; $C = 0.004064$.

4.3.4 Thermal Diffusivity

Thermal diffusivity is an intensive wood property defined as the ratio of thermal conductivity to the wood density and average isobaric specific heat:

$$\alpha = k/\rho \cdot \bar{c}_p \tag{4.6}$$

where
α is the thermal diffusivity $\left(\mathrm{m}^2/\mathrm{s}\right)$;
k is the thermal conductivity $(\mathrm{W/m \cdot K})$;
ρ is the wood mass density $(\mathrm{kg/m^3})$; and
$\bar{c}_p$ is the wood average specific heat at constant pressure $(\mathrm{J/kg \cdot K})$.

Thermal diffusivity of softwood and hardwood varies with moisture content, as shown in Figure 4.2.

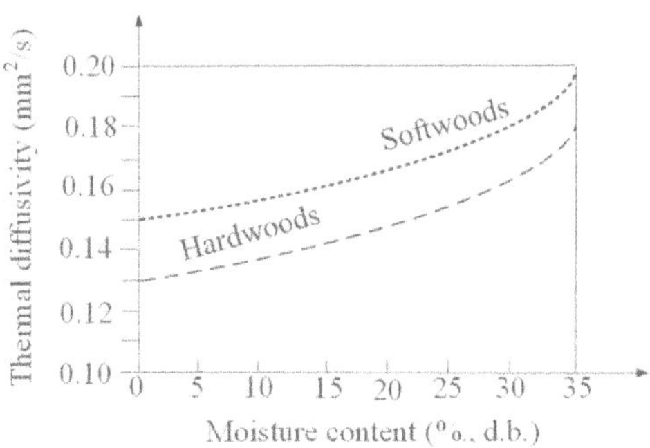

FIGURE 4.2 Typical variation of thermal diffusivity of softwood (e.g., spruce and black pine) and hardwood (e.g., oak, chestnut, and beech).

4.3.5 Moisture Content

Moisture has a significant influence on green timber and sawn logs, as well as on lumber boards because it continually changes according to the ambient (e.g., temperature and relative humidity) conditions.

Moisture in wood is present in the following forms (Skaar 1988): (i) free water in the cell cavities (mainly diffused by capillary tension with high drying rates), (ii) water vapor within the cell walls, and (ii) unbound water in voids.

Moisture content of a given piece of wood is usually expressed as a percentage and can be calculated as follows:

$$MC = (m_{water} / m_{oven-dry}) \times 100\,(\%) = \left[\left(m_{wet} - m_{oven-dry}\right) / m_{oven-dry}\right] \times 100\,(\%) \quad (4.7)$$

where

MC is the moisture content ($kg_{water} / kg_{oven-dry}$);
m_{water} is the mass of water in the wood specimen (kg_{water});
$m_{oven-dry}$ is the mass of oven-dried wood specimen ($kg_{oven-dry}$); and
m_{wet} is the mass of the specimen at a given moisture content (kg_{wet}).

Table 4.3 shows the average moisture contents of heartwood and sapwood of some green hardwood and softwood species.

In conventional hot air convective batch dryers, drying times of hardwood (as yellow birch) and those of softwood (as white spruce) could be as long as 200 hours and 70 hours, respectively (Table 4.4).

TABLE 4.3
Average Moisture Contents of Some Green Hardwood and Softwood Species

Species	Moisture Content (%)
Hardwood	
White birch	74 (d.b.) (heartwood); 92 (d.b.) (sapwood)
Hard maple	30–41 (d.b.)
American elm	95 (d.b.) (heartwood); 92 (d.b.) (sapwood)
Maple sugar	65 (d.b.) (heartwood); 72 (d.b.) (sapwood)
Willow oak	64 (d.b.) (heartwood); 78 (d.b.) (sapwood)
Yellow poplar	83 (d.b.) (heartwood); 106 (d.b.) (sapwood)
Softwood	
Green black spruce	52 (d.b.) (heartwood); 113 (d.b.) (sapwood)
Green balsam fir	88 (d.b.) (heartwood); 173 (d.b.) (sapwood)
Pine, western white	62 (d.b.) (heartwood); 148 (d.b.) (sapwood)
Cedar, western red	58 (d.b.) (heartwood); 249 (d.b.) (sapwood)
Cypress, black	121 (d.b.) (heartwood); 171 (d.b.) (sapwood)
Redwood	86 (d.b.) (heartwood); 210 (d.b.) (sapwood)

Note: d.b., dry basis.

TABLE 4.4
Approximate Drying Times of Some Wood Species

Dried Product	Drying Time (hours)
Hardwood	
Yellow birch	≈200
Softwood	
White spruce	≈65–70

In the case of hard maple (hardwood), the drying time could be 120 hours when dried at dry-bulb temperatures between 71°C and 82°C (Figure 4.3).

After drying, the overall final moisture contents of hardwood lumbers (used for furniture, cabinets, and millwork) generally range around 7.5% (d.b.) (Table 4.5).

Softwood lumbers intended for framing in construction are usually dried to an average moisture content of 15%, not to exceed 19% (dry basis). For other uses, as for appearance grades, softwood lumbers are dried to moisture contents of 10–12% (dry basis), and to as low as 7–9% (dry basis) for furniture, cabinets, and millwork.

The most used measurement methods of wood moisture contents are (Harley and Marchant 1995) (i) oven-dry, (ii) electric or capacitive meters, and (iii) lumber stack(s) weighting prior, during, and after drying.

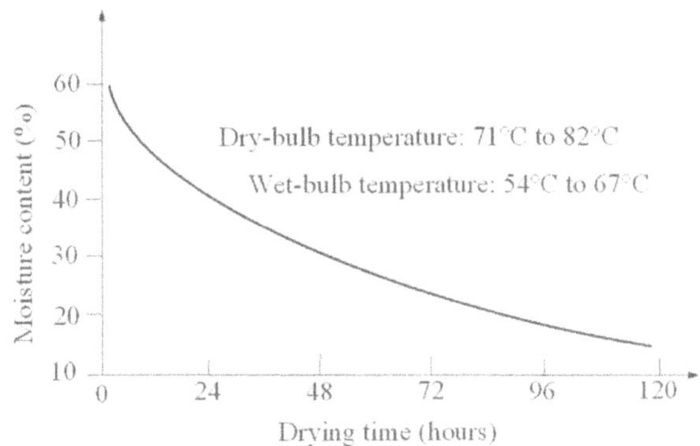

FIGURE 4.3 Variation of moisture content (d.b.) with drying time of hard maple (hardwood).

TABLE 4.5
Average Final Moisture Contents of Two Dried Hardwoods

Species	Moisture Content (%)
Yellow birch	7.5 (d.b.)
Hard maple	7.5 (d.b.)

Note: d.b., dry bulb basis.

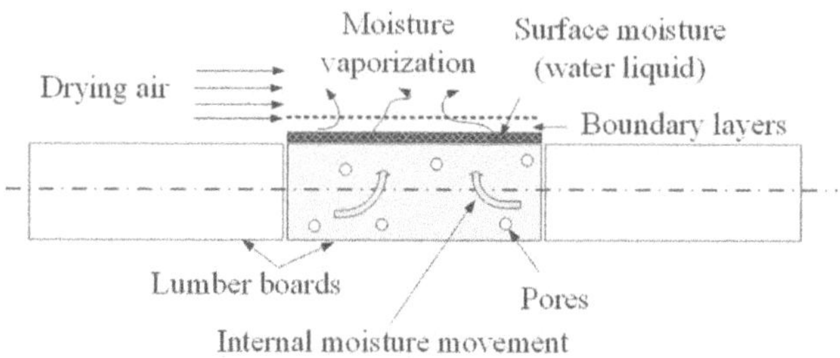

FIGURE 4.4 Schematic of internal and external (surface) drying processes. *Note:* Schematic not to scale; not all processes are shown.

4.3.6 Moisture Diffusivity

Wood is a hygroscopic material that reacts to changes of moisture content, mainly due to changes in ambient air temperature and relative humidity.

Movement of moisture (water) through dried woods can be explained by mechanisms such as (i) diffusion of liquid caused by concentration gradients, (ii) partial vapor pressure, (iii) capillary forces, (iv) gravity forces, and (v) surface diffusion.

Moisture diffusivity in wood refers to the rate at which moisture molecules move through the wood matrix (Figure 4.4). It is influenced by the material moisture content and temperature and becomes stronger below the fiber saturation point (Franke et al. 2016).

The rate of moisture movement in woods is described by a parameter called effective moisture diffusivity (D_{eff}), of which values vary due to factors such as material properties and structure and drying parameters (e.g., temperature).

Figure 4.5 shows a typical variation of wood moisture diffusivity with moisture content at drying temperature of 60°C.

Fick's first law relates the moisture diffusive mass flux to the gradient of the moisture content. It postulates that the moisture mass flux goes from regions of high moisture contents to regions of low moisture contents, and that its magnitude is proportional to the moisture content gradient (spatial derivative):

$$\dot{m}_{moisture} = -D_{diff} \cdot (d\varphi / dy) \tag{4.8}$$

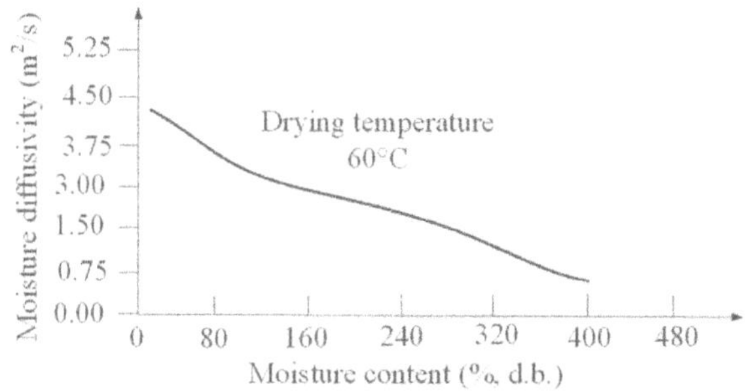

FIGURE 4.5 Typical variation of wood moisture diffusivity with moisture content at drying temperature of 60°C.

where

$\dot{m}_{moisture}$ is the specific moisture mass diffusion flux (also called density of moisture mass diffusion) that measures the amount of moisture that flows through one unit area during one unit time interval ($kg_{water}/m^2 \cdot s$);

D_{diff} is the moisture diffusion coefficient (diffusivity), which depends on the material particle sizes, moisture content, and temperature (m^2/s);

φ is the amount of moisture (particles) concentration per unit volume (kg_{water}/m^3);

$d\varphi$ is the gradient of moisture (particles) concentration (kg_{water}/m^3);

dy is the change in moisture particles' positions (m); and

$d\varphi/dy$ is the linear gradient of moisture concetration per unit volume; this is the driving force for one-dimensional moisture diffusion that occurs in response to moisture content gradients expressed as the change in moisture content due to a change in position ($kg_{water}/m^3/m$).

The negative sign in Equation 4.8 indicates that moisture diffusion occurs in an opposite direction to that of the increasing of moisture content. Hence, diffusion occurs in the direction of decreasing concentration of the diffusing moisture, and thus, the moisture diffusion flux is a positive quantity.

For one-dimensional dried wood lumber boards, transient moisture diffusion is defined through Fick's second law that predicts the changes (gradients) in moisture particles' mass concentrations with time (Crank 1975; Franke et al. 2016).

4.3.7 Moisture Sorption Isotherms

Sorption (adsorption and desorption) isotherms are graphs of moisture content measured experimentally under isothermal conditions.

Adsorption isotherms are obtained by exposing the wood specimen to the air of increasing relative humidity, while desorption isotherms are obtained by exposing the wood to air of decreasing relative humidity at constant temperature (Figure 4.6a).

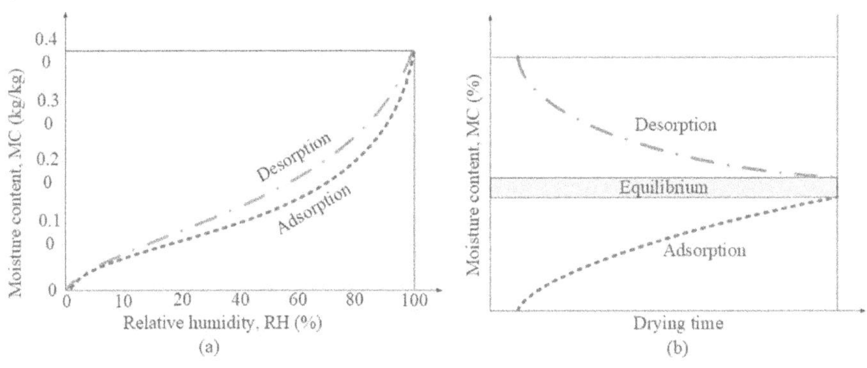

FIGURE 4.6 Typical representations of adsorption and desorption isotherms as functions of (a) relative humidity of drying air; (b) drying time.

It can be seen that moisture (adsorption and desorption) isotherms are nonidentical and nonreversible, a property called sorption hysteresis. It can also be seen (Figure 4.6b) that the EMC is lower during desorption and higher during adsorption (Glass and Zelinka 2010).

4.3.8 Fiber Saturation Point

In the case of dried wood, fiber saturation point (FSP) is defined as the moisture content at which all free water has been removed but the cell walls are still fully saturated. This occurs at moisture content of approximately 30% (d.b.) for most of wood species (Keey et al. 2000).

Fiber saturation point is an important property of wood drying due to the following facts: (i) below FSP, more energy is needed to remove the bound water than liquid water; (ii) wood starts to shrink with moisture content reduction below the fiber saturation point, and vice versa; and (iii) wood mechanical properties (e.g., strength) increase with the decrease in moisture content below the fiber saturation point.

For all species of woods, the moisture content at fiber saturation point can be expressed as a temperature function by the following empirical equation (Siau 1984):

$$MC_{FSP} = 0.30 - 0.001 \cdot \left(\bar{T} - 20\right) \tag{4.9}$$

where

MC_{FSP} is the moisture content at fiber saturation point ($kg_{water} / kg_{oven-dry}$); and
$\bar{T}$ is the average drying temperature (°C).

At moisture content above the fiber saturation point, moisture movement within the wood consists of transfer of bound, liquid, and water vapor phases through the cell walls and the interconnected voids. Above the fiber saturation point there is practically no change in wood volume (shrinkage or swelling) with changes in moisture content (Arends et al. 2018; Krabbenhoft 2003).

Below the fiber saturation point, bound water diffusion is dominant and many properties of wood considerably change as follows (Desch and Dinwoodie 1996): (i) shrinkage begins since the moisture removed comes from the cell walls; (ii) mechanical properties such as strength, hardness, and toughness increase; (iii) rapid increase of wood electric resistance because the loss of bound water; (iv) decrease of drying rate as moisture movement of bound water is slow compared to unbound water; and (v) more energy is required to vaporize (or evaporate) the bound water because the attraction between wood and water must be overcome.

4.3.9 EMC

Living trees contain large quantities of water. They adjust their moisture content to the surrounding conditions (e.g., relative humidity and temperature) by the adsorption or desorption and diffusion of water vapor until the EMC is reached.

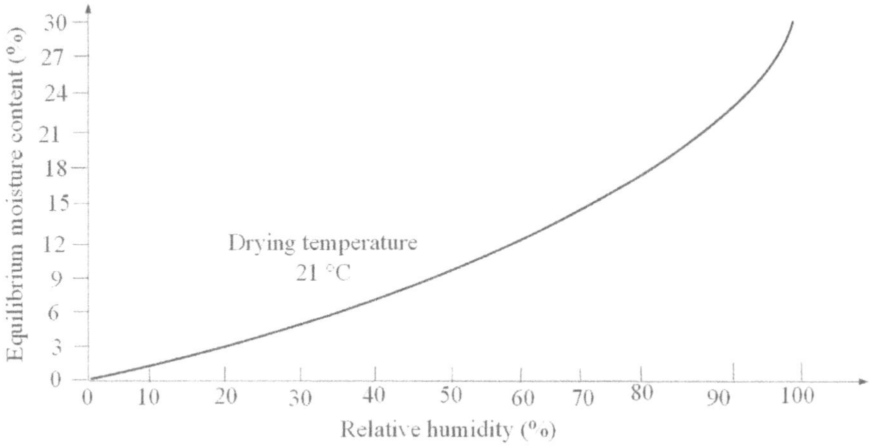

FIGURE 4.7 EMC of wood as a function of relative humidity of surrounding air at 21°C. (Redrawn from Krabbenhoft 2003.)

Wood EMC is approximately proportional to the ambient relative humidity (a function of temperature), and, to a lesser degree, with temperature. It increases until the relative humidity of ambient air approaches 100% (Siau 1984; Krabbenhoft 2003) (Figure 4.7).

4.4 ADVANTAGES OF DRYING WOOD

Due to its wide availability and renewability, wood is a highly durable material having applications mostly in building construction, cabinetry, paneling, furniture, and high-grade flooring material. However, the fibrous and hygroscopic mass of freshly felled trees and recently cut green timbers contain up to 75% of water, which affects the wood's physical properties as dimension, strength, and hardness.

Therefore, the sawn timber should be dried to remove the moisture (water) to certain predefined levels, usually, below 20% (d.b.) before being processed into end-use products. Wood products to be used inside heated buildings (homes, offices) that are only occasionally heated should be dried to about 6–8% moisture content (Derome et al. 2015).

The dried lumber has the following advantages over the undried green timber (Derome et al. 2015): (i) prevents wood degradation after felling; (ii) reduces handling, storage, and transportation costs; (iii) dried lumber works, machines, finishes, and glues are better than green timber; (iv) increases the wood's strength, making it more compatible with the normal service conditions; (v) improves mechanical and structural support of buildings and objects as furniture; (vi) improves dimensional stability for various finishing processes, such as polishing and painting, while paints last longer on dry timber; (vii) improves thermal insulation properties of wood; and (vii) adds value to sawn timber materials, sometimes up to 700%, the value to raw timber.

4.5 CONVENTIONAL DRYING METHODS OF WOODS

The principal scope of wood drying methods is to remove the moisture (water) from the sawn timber at the lowest possible costs and without compromising its final commercial value.

This section briefly presents some of traditional techniques used for wood drying, such as natural (yard) air, hot air convective kiln, and vacuum and superheated steam dryers. Once the appropriate drying method is selected, it is important to choose the optimum operating conditions to provide the required lumber quality at lower energetic costs.

4.5.1 Natural (Yard) Air Drying

Natural (yard) air drying of large stacks of lumber is performed by exposing to atmospheric conditions. This weather-dependent drying process is cost-effective since it uses natural, free energy supplied by ambient air, sun, and wind.

Pre-drying of lumber stacks is performed in sawmill yards until wood moisture content drops, depending on species, thickness, location, and the time of year the lumber stacked, at between 20% and 30% (d.b.) ahead of forced hot air from convective kilns (Mujumdar 2015). Another objective is to reduce the chances of the wood to develop mold, stain, or decay during transit and storage. Table 4.6 shows some approximate air drying times for freshly cut (green) 25 mm thick lumber boards to attain 20% (d.b.) moisture content (Walker et al. 2013).

Low-density lumber species (such as pine, spruce, and soft maple), stacked in favorable locations and times of the year, may require 15–30 days for air drying, while slow-drying species (such as red and white oaks) in northern locations and stacked at unfavorable periods of the year could require up to 300 days. However, southern pine and yellow poplar will dry much faster in the yard.

The following are the advantages of natural (yard) air drying of wood (Cech and Pfaff 2000): (i) relatively low initial capital and operating costs, (ii) massive reduction of fossil energy consumptions, and (iii) reduction of the subsequent hot air convective kiln capacity and drying time.

The following are the disadvantages of natural air drying : (i) the output is dependent on the local climate (ambient dry- and wet-bulb temperatures, relative humidity, wind velocity, rainfall, sunshine, etc.); (ii) depending on the weather conditions, the natural air drying process may take several months, or years; and (iii) lack of absolute control of drying conditions.

TABLE 4.6
Air Drying Times Up to 20% (d.b.) for 25 mm Thick Lumber

Species	Days
Red oak	100–300
White oak	150–300
Southern pine	40–150
Yellow poplar	60–150

In order to increase the wood drying rate and decrease the drying times, stickered packages of lumber boards are sometimes placed in unheated, permanent roofed structures (to protect the lumber from rain and direct solar insolation) having open sides. To accelerate the drying process, the atmospheric air is forced through the wood piles at velocities of 2.5–3.6 m/s by multiblade fans.

4.5.2 Hot Air Convective Batch Kiln Drying

Convective hot air kiln drying (Figure 4.8) is a cost-effective wood drying method commonly used in the wood industry worldwide.

The aim of this drying method is to lower the initial moisture content of wood lumber boards from 20–60% (d.b.) to moisture contents of 6–12% (d.b.) before they can be stored and transported.

In cold climate regions, timber drying is the primary consumer of energy at sawmills (80% of the sawmill's total energy consumption) in the form of heat 80–90% and electricity for air circulation fans 10–20%, while significant amount of the input energy is vented away in the form of hot and moist air.

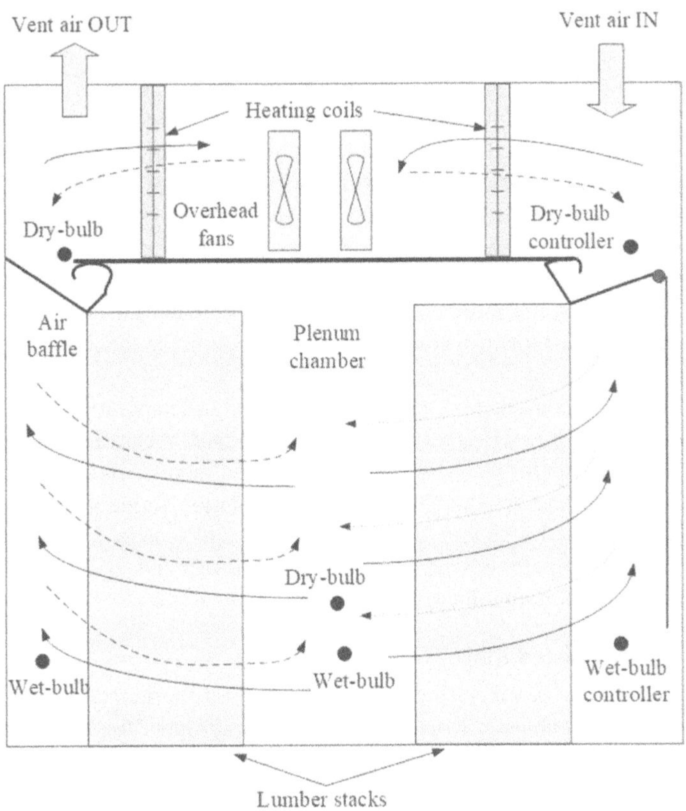

FIGURE 4.8 Schematic of a conventional hot air convective wood batch dryer (kiln).

In most conventional hot air convective industrial batch kilns, heat is supplied externally to provide energy for internal moisture movement and evaporation.

Compared to the natural (yard) air drying, conventional hot air convective kiln drying has the following advantages (Cech and Pfaff 2000): (i) more efficient drying method, (ii) allows better control of the final moisture content regardless of weather conditions, (iii) reduced the drying time;, and (iv) ensures higher throughput.

The following are the advantages of conventional hot air convective kiln drying : (i) higher capital, operation, and maintenance costs, and (ii) heat losses and inefficiencies in heat transfer usually limit the drying dry-bulb temperatures to maximum 82–93°C.

Most drying kilns operate at low (50–60°C) and medium (60–80°C) temperatures. However, it is possible to run the drying process at much higher temperatures (80–120°C) and even up to 160°C, allowing the moisture (water) to migrate and vaporize by boiling off, still at atmospheric pressure.

Concerning the kiln wood drying at high temperatures (80–120°C), the following aspects could be mentioned (Minea 2019): (i) subject to higher energy losses, (ii) shorter drying times may reduce the heat losses by air leaks and overhead fans' electricity consumption up to 20% over long terms, (iii) drying capacities could be more than 100% higher when the drying temperature is raised from 70°C to 90°C and almost 400% higher when the temperature is raised to 115°C, (v) total drying costs could be lower, and (v) lumber quality after drying could be not significantly deteriorated.

The following are the limitations of high-temperature wood drying methods compared with those performed at low- and medium temperatures (Hajian et al. 2024): (i) require more expensive initial investments for equipment and kiln construction; (ii) the moisture evaporation rate is low in wood species with low permeability, which may cause pressure buildup, nonuniform moisture content distribution, internal stresses, and, consequently, more drying defects (such as distortions and deformations); and (iii) require more conditioning, usually done by steaming or spraying water, to equalize the moisture gradient and internal stress at the end of each drying process.

In addition, the construction and operation of high-temperature kilns requires special measures such as (i) higher construction standards regarding tightness, stainless steel or aluminum for walls, doors, and roofs, fan motors, and buffers; (ii) more thermally efficient and resistant to corrosion insulation (mineral or glass wool); (iii) conventional psychrometers can be used as special sensors for process control since the wet bulb temperatures are generally below 100°C; and (iv) indoor air velocity should be adapted to maintain uniform drying.

4.5.2.1 Solar-Assisted Kilns

Daily world-average solar radiation (insolation) on horizontal surfaces is of 3.82 kWh/m^2, with values in tropical countries being higher (up to 7.15 kWh/m^2). For drying of lumber, solar-assisted dryers could be attractive in tropical or subtropical climates and remote locations spread over latitudes from 0° to 50° (Keey et al. 2000).

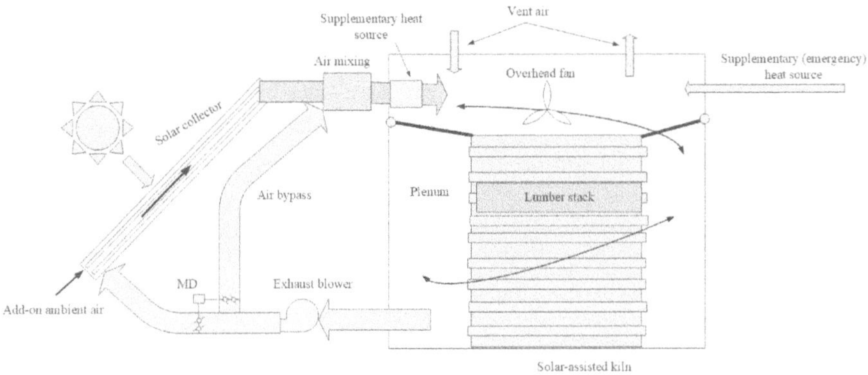

FIGURE 4.9 Simplified representation of a typical conventional solar-assisted lumber batch kiln. MD, motorized dampers.

In solar-assisted kilns (Figure 4.9), the thermal solar collectors, constructed of flat or corrugated metal sheets painted in black, filled with either air or water, can be installed on separated structures or on the dryer roof. Exhaust blowers circulate air through the solar collectors and through the lumber stack(s). Inside the drying chamber, the air circulation is forced by overhead fans. Air venting can be provided by pressure gradients or via powered fans controlled by humidity, temperature, and/or external conditions such as sunlight.

In such drying systems (Langrish et al. 1992) (i) supplemental energy (e.g., from oil, natural gas or electricity) is necessary, (ii) solar thermal captors should be isolated from the dryer during night hours and during periods of low solar influx, (iii) the choice of solar thermal captor surfaces in practice is restricted to roof and south wall, and (iv) when the kiln's air vents close overnight and with the drop in outdoor temperature, the air relative humidity inside the kiln rises sufficiently for moisture to condense on the wood boards' surfaces; this may provide a degree of conditioning, which prevents the development of lumber excessive checking.

The following are the disadvantages of solar-assisted dryers: (i) intermittent solar energy source, depending on time, season, and weather; (ii) relatively low-energy flux compared with conventional energy sources; (iii) higher capital costs are incurred for additional solar panels, blowers, and valves; and (iv) the amount of stored solar energy (if required) is greatly subjected to the weather conditions.

4.5.2.2 Direct Fired Kilns

Direct firing kilns are frequently used in wood drying industry (Figure 4.10), mainly because this drying technology is more rapid and more cost-effective compared, for example, with electric heating methods.

The use of direct fired burners inside the wood dryers is useful especially in winter to melt snow and ice ensuring an efficient and rapid rise of air temperature inside the wood dryer. Overhead fans circulate the warm air across the lumber stacks ensuring uniform air distribution in the dryer.

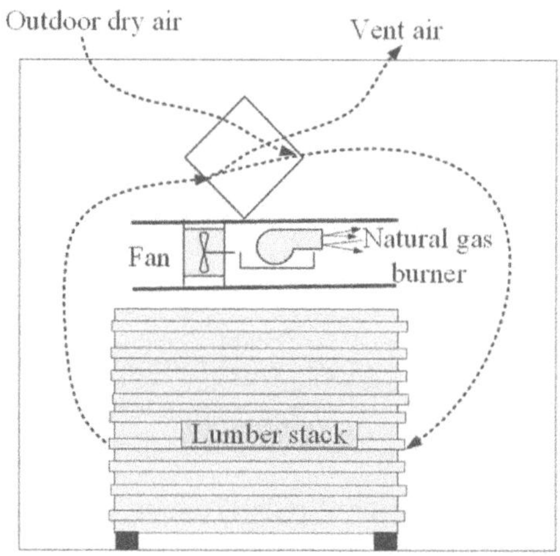

FIGURE 4.10 Schematic of a direct fired wood kiln. *Note:* Schematic not to scale; not all components are shown.

Drying medium is the combustion gases from direct-fired burners using natural gas (with calorific thermal energy ranging from approximately 37–45 MJ/m^3, including the condensing latent heat of water vapor formed by combustion), oil, diesel, liquefied propane, or butane as energy sources, depending on their relative local prices. In the heating chamber, the hot combustion gases are mixed with the circulating air, raising its temperature. Centrifugal blowers and ducts distribute the hot air to the plenum chamber.

The following are the advantages of direct fired kilns (Cech and Pfaff 2000): (i) ease of installation; (ii) uniform heating and drying; (iii) better control of temperature; (iv) reduced maintenance; (v) potential annual savings of up to 10% compared with indirect heating dryers with heat exchangers; since latent heat also serves to humidify the wood stack, drying efficiency could reach levels as high as 95%.

The following are some of the limitations of this wood drying technique: (i) important energy losses through the exhausted combustion flue and air mixture, and (ii) expensive heat recovery devices from exhausted air.

4.5.3 Vacuum Drying

Vacuum drying is generally used for relatively small production of wood heat-sensitive products that must be dried at low temperatures over relatively short periods of time. This technique requires sealed enclosures (Figure 4.11), where the air is evacuated to provide a very low air pressure (e.g., 0.02 MPa at 80°C). In such vacuum conditions, moisture (water) can evaporate at much lower temperatures.

The energy used to evaporate the moisture (water) can be supplied by (i) conduction (with aluminum platens heated by electrical resistance or circulation of hot water placed between each layer of boards, (ii) radiation, and (iii) forced hot air.

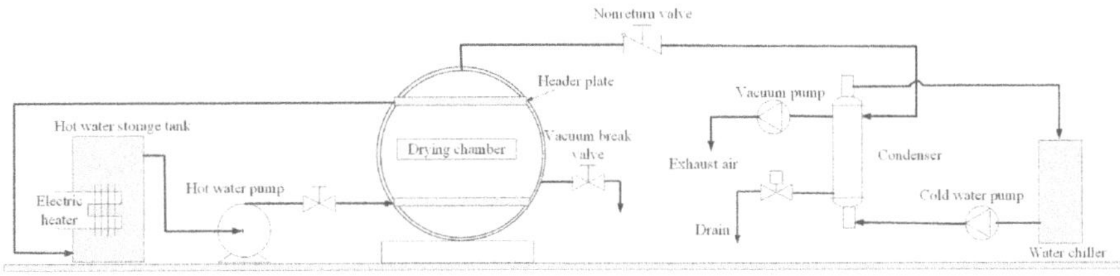

FIGURE 4.11 Schematic of a wood vacuum drying system.

The discontinuous vacuum drying kilns consist of repeated cycles of heating under vacuum followed by cooling periods (each of about one hour) during which the evaporated moisture is condensed on the cooled platens, recovering some of the moisture, and removed from the drying enclosure.

The following are the advantages of vacuum dryers for drying thick lumber boards (Hilderbrand 1989; Walker et al. 2013): (i) reduces drying times by up to 50% compared with conventional kilns at atmospheric pressure, (ii) accelerates the drying rates, (iii) provides good quality (e.g., color preservation, avoiding discolorations, and better control of volatile organic compound emissions) of heat-sensitive wood species, (iv) compared to drying at atmospheric pressure, vacuum drying reduces the vapor pressure and allows moisture to evaporate faster at the same drying temperature, (v) commercial vacuum drying operations may achieve drying rates 3–17 times faster than ambient pressure drying, (v) no danger of developing defects with some susceptible species, and (vii) may reduce the (electrical and fossil) energy consumption by up to 70%.

The following are the disadvantages of vacuum drying (Keey et al. 2000): (i) capacity of kilns is relatively low compared to the cost of equipment, (ii) cost of electrical energy is usually high in many countries, and (iii) high initial capital since almost all components of kilns must be made from stainless steel or aluminum.

4.5.4 Superheated Steam Drying

Superheated steam drying of wood is a convective drying technique compared to hot air convective kiln drying, but with the main difference is that only pure steam (water vapor at above its saturation temperature) is used as the drying medium in place of hot air, or combustion gases. It is a justifiable option only for very large and continuously operated drying facilities.

Superheated steam kilns (Figure 4.12) can operate at low pressure (vacuum), near-atmospheric pressure, or higher pressures. For temperatures below 100°C, the saturation pressure of steam is below atmospheric pressure, thus the dryer needs to be operated in vacuum. For that, air pressure of the drying chamber is lowered by evacuating air with a vacuum pump, followed by steam being introduced into the drying chamber. At pressures below atmospheric pressure, steam would be in superheated phase at lower temperatures, usually in the range of 50–90°C.

The following are the advantages of superheat steam drying (Cech and Pfaff 2000): (i) higher drying rates; (ii) better final quality for certain wood species,

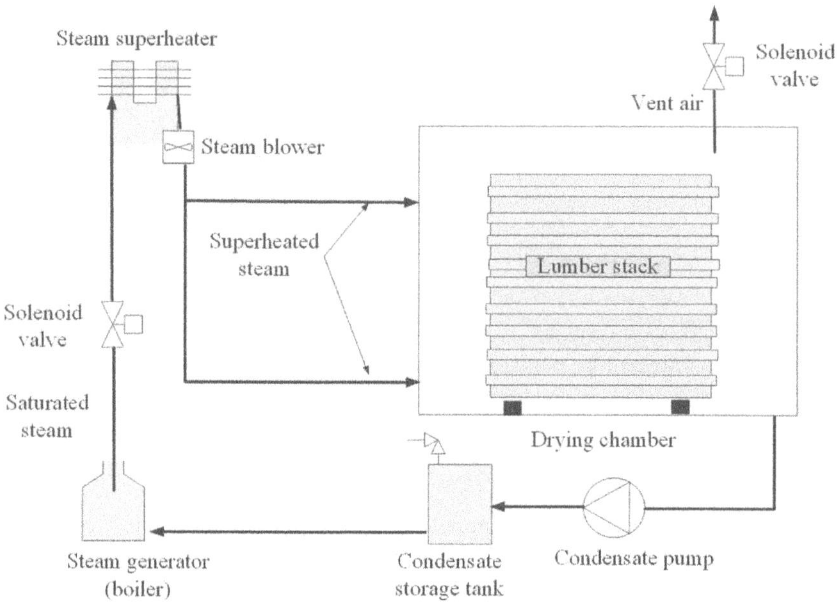

FIGURE 4.12 Schematic diagram of a superheated steam wood kiln.

because of staining, mold attack can be avoided, and no oxidative discoloration of wood may occur; (iii) lower net energy consumption, mainly because there is no need for fresh air ventilation and humid air exiting; (iv) no fire and explosion hazards; (v) higher production capacity resulting in energy savings of about 50% over conventional kilns; (vi) faster drying rates going up to less than seven days; (vii) the vapor evolving from the dried wood lumber may be withdrawn from the drying chamber, condensed and, thus, the latent heat recovered up to 90%; (viii) reduced wood stresses, cracks, and warps; and (ix) up to 50% lower energy consumption and up to 80% reduction in drying time compared with hot air convective drying.

The following are the limitations of superheated steam drying (Mujumdar 1995; Walker et al. 2013): (i) more complex drying concept (because of start-up and shutdown rules); (ii) higher initial capital costs; (iii) steam leaks are not allowed; and (iv) feeding and discharge of superheated steam must not allow infiltration of non-condensable gases (as air).

4.6 TYPICAL WOOD DRYING CURVES

Typical wood drying curves can be represented by a series of drying periods, each characterized by different dry- and wet-bulb temperatures, moisture content levels, and drying rates. The drying process is influenced by factors such as (i) species of dried wood, (ii) environmental conditions, and (iii) method of drying. For example, yard (natural air) drying is slower and can take several months or even years, while kiln drying uses controlled heat and humidity to accelerate the drying process. Wood drying curves can be predicted using overall moisture (mass) transfer coefficients,

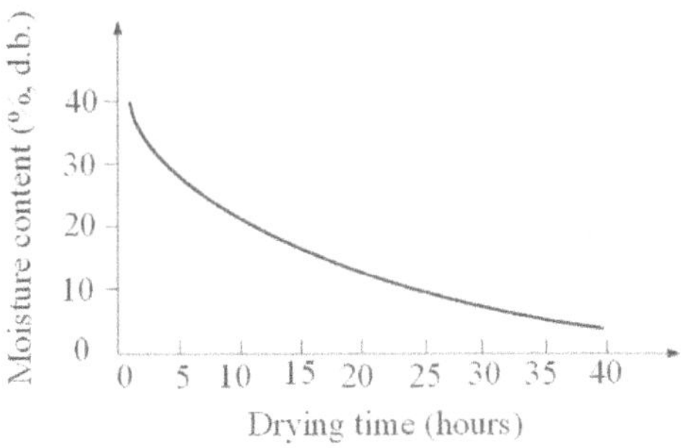

FIGURE 4.13 A typical wood drying curve plotted as variation of average moisture content versus drying time in the case of conventional hot air convective batch dryers.

which accounts for both internal and external transfer resistances during the drying processes.

In both convective and heat pump-assisted wood drying systems, to maintain approximately constant drying rates, the material structure and water molecules must be supplied with additional energy (heat) or the partial vapor pressure of the drying air must be decreased either by raising the temperature or reducing its relative humidity. The moisture (water) evaporation rate from wood stacks depends on the amount of heat supplied and the capacity of the drying air to absorb it.

Wood drying curves usually plot the variation of average moisture content versus drying time, as shown in Figure 4.13.

In the case of heat pump-assisted wood drying systems, the total mass of moisture extracted (and condensed) by the heat pumps' evaporators (that strongly depends on the wood species and initial and final moisture content) can be visualized, as shown in Figure 4.14 (Minea and Normand 2005).

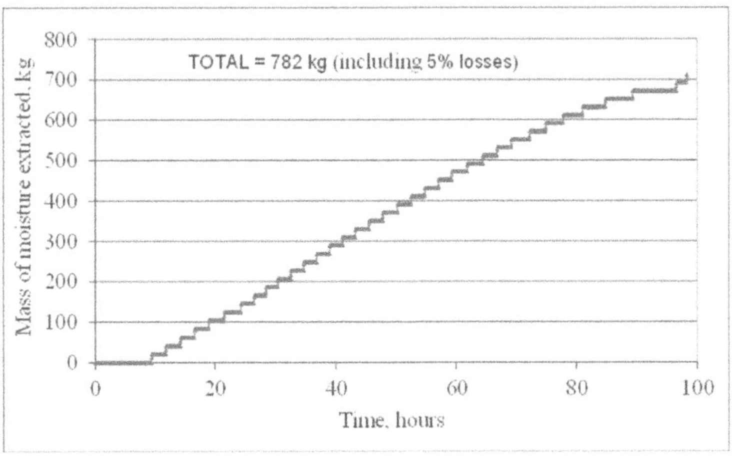

FIGURE 4.14 Cumulative mass of moisture extracted, including 5% of condensate losses (heat pump-assisted softwood drying). (From Minea and Normand 2005.)

4.7 TYPICAL WOOD DRYING SCHEDULES

Typical wood drying schedules are developed to control the time changes of drying air dry- and wet-bulb temperatures, and relative and absolute humidity in accordance with the moisture content and stress phenomena within dried lumber boards (Perré and Keey 2015). In many countries and/or geographic regions, typical (generic) kiln drying schedules (also called indexes or programs) have been developed based on empirical trial-and-error studies from small-scale tests involving changes of several pertinent parameters.

The following basic points should be considered when a typical wood drying schedule is selected: (i) type of kiln; (ii) wood species (because there are many variations in anatomical, physical, and mechanical properties between them); (iii) width and thickness of lumber boards (because the drying time is approximately inversely related to thickness and, to some extent, influenced by the width of dried pieces); (iv) whether the lumber boards are quarter-, flat-, or mixed-sawn (because sawing pattern influences the distortion due to shrinkage anisotropy); (v) permissible drying degrade (because aggressive drying schedules can cause timber to crack and distort); and (vi) intended use of timber (because the required appearance of the timber surface and the target final moisture contents are different depending on the uses of timber).

Figure 4.15 shows the main stages of a typical wood drying schedule.

During the preheating (warming-up) step, generally performed at dry-bulb temperature of 95°C, lumber boards are heated to a point where the drying air relative humidity reaches 100%, which improves energy transfer, plasticizes the material and, thus, restricts stress development, minimizes subsequent degradations (such as checking or honeycomb), and destroys the microorganisms responsible for sapwood discoloration.

During the capillary diffusion step, moisture (liquid water) evaporates as drying air's relative humidity decreases, and the gap between dry-bulb and wet-bulb temperatures (called wet-bulb depression) increases. In transition regime, most of the liquid water evaporates, while part of it is present in isolated regions. Then, bound water present in cell walls evaporates. During bound water diffusion step, only the bound water is removed by forced evaporation, and the lowest relative humidity

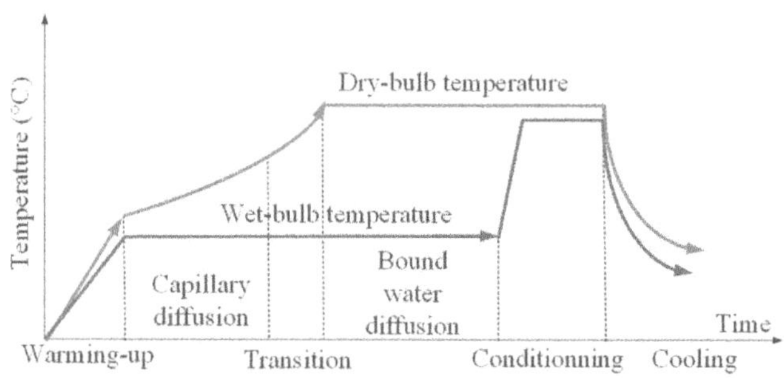

FIGURE 4.15 Typical (generic) stages of a wood drying cycle.

TABLE 4.7
Example of Drying Schedule Index for Eastern Canadian Softwood Species

	Spruce		Jack Pine and Balsam Fir	
Moisture Content% (d.b.)	Dry-Bulb Temperature °C	Wet-Bulb Temperature °C	Dry-Bulb Temperature °C	Wet-Bulb Temperature °C
>40	71.0	66.0	N/A	N/A
40 to 30	76.5	65.5	N/A	N/A
>35	N/A	N/A	82.2	79.4
35 to 25	N/A	N/A	87.7	79.4
<30	82.2	54.4	N/A	N/A
<25	N/A	N/A	93.3	65.5

Note: d.b., dry bulb.

of the drying air is reached. Conditioning step consists in abruptly increasing the relative humidity to eliminate differences in moisture content gradients in lumber boards and reduce drying-induced stresses that may cause distortions of wood in use. Finally, the cooling step is performed when lumber boards are removed from warm drying enclosures and cooled in a controlled way to avoid cracks and uncontrolled further drying.

Table 4.7 shows, as an example, a schedule index for Eastern Canadian softwood species as spruce, jack pine, and balsam fir (Cech and Pfaff 2000).

Typical drying indexes (as shown in Table 4.7) are usually revised (adapted) to the actual drying processes (e.g., to heat pump-assisted wood drying) according to the quality, board dimensions, initial temperature, and moisture content of dried species. In other words, most of actual drying schedules used throughout the industry are based on generic (regional) index schedules adapted (modified) by sawmills, however, without major changes compared to the conservative schedules developed in the past.

Table 4.8 shows such a revised (actual) wood drying schedule for a heat pump-assisted wood drying cycle. It can be seen that each cycle begins with a preheating (warming-up) step (when the heat pump is not allowed to run), with set dry-bulb temperatures between 87.7°C and 93.3°C, and set wet-bulb temperatures between 68.3°C and 79.4°C, in order to destroy the stain producing organisms (Minea and Normand 2005).

In this example, each of the following five steps has been set for maximum 10 hours, and the sixth for a maximum of 20 hours (Minea 2004).

4.8 WOOD DRYING DEFECTS

In drying of most softwoods and some hardwoods, free water movement is driven by liquid pressure gradient due to capillary actions, whereas bound water movement is by diffusion.

TABLE 4.8
Example of an Actual Wood Drying Schedule

	Preheating			Step 1			Step 2		
Test Number	Set DB °C	Set WB °C	Actual Time hours	Set DB °C	Set WB °C	Set Time hours	Set DB °C	Set WB °C	Set Time hours
#1- *CONV*	93.3	62.7	–	79.4	68.3	10	79.4	65.5	10
#2	87.7	68.3	5.03	76.6	62.7	10	79.4	62.7	10
#3	87.7	68.3	5.56	76.6	62.7	10	79.4	62.7	10
#5	87.7	73.8	8.28	79.4	62.7	10	79.4	62.7	10
#6	93.3	79.4	8.03	85.0	155	10	85.0	68.3	10

	Step 3			Step 4			Step 5		
Test Number	Set DB °C	Set WB °C	Set Time hours	Set DB °C	Set WB °C	Set Time hours	Set DB °C	Set WB °C	Set Time hours
#1- *CONV*	79.4	64.4	10	82.2	64.4	10	82.2	62.7	20
#2	79.4	62.7	10	82.2	62.7	10	82.2	60.0	20
#3	79.4	62.7	10	82.2	62.7	10	82.2	60.0	20
#5	79.4	62.7	10	82.2	62.7	10	82.2	60.0	10
#6	87.7	71.1	10	90.5	71.1	10	93.3	71.1	20

Note: DB, dry-bulb temperature; WB, wet-bulb temperature.

Proper drying of wood under controlled conditions prior to use is of great importance since if it is inappropriately dried (e.g., too fast or too slow), one or more drying irreversible defects may occur: (i) shrinkage, (ii) splits (separation of wood fibers along the grain forming cracks or fissures), (iii) checks (lengthwise separation of wood fibers that extend across the annual growth rings), (iv) warps/distortions causing departure from their original planes, (v) casehardening (a condition of varying degrees of stress set in wood such that the outer wood fibers are under compressive stress and the inner fibers under tensile stress), (vi) honeycomb (checks) that occurs most often in the interior of the wood, usually along the wood rays, and (vii) collapse (resulting from the physical flattening of fibers above the fiber saturation point).

Too fast drying would generate large moisture content gradients, which are likely to induce high drying stresses and occurrence of wood collapses for many species of wood.

On the other hand, too slow drying would lengthen the drying time, increase energy consumption, and reduce productivity.

4.8.1 Shrinkage

Porous solid materials such as wood have high initial moisture content and porosity. With respect to dimensional stability, wood is an anisotropic material. During drying, lumber boards exchange moisture with the drying air according to the level of its relative humidity. Too rapid decrease of moisture content may cause shrinkage, a process defined as total variation of dimension between the green state and the oven-dry condition. The evaporation of water during the drying process is often accompanied by significant volume changes and tissue shrinkage. The deformation in the material drying process directly affects the heat and mass transfer pattern of the material, which in turn affects the drying efficiency, the energy consumption of the drying process, and the final product quality.

Moisture content at the surface or in the near-surface layer is lower than that in the inner layers. When the local moisture content in the surface layers decreases to below the fiber saturation point, these layers start to shrink, whereas the inner layers have higher moisture content and have not yet started shrinking. As the overall thickness of the surface layer is much less than the inner layers, the tensile stress in the surface layer may be greater than the maximum elastic limit, in which case permanent strain is formed. Once the permanent strain is formed, the wood cannot be recovered to its original state even if the stress is relieved.

Migration of bound water within the tissue during drying has a significant effect on the shrinkage of the material. The amount of shrinkage of dried wood increases with the increase in dry-bulb temperature.

This phenomenon is influenced by factors such as the structure of wood cells and the drying conditions.

Shrinkage (volume change) of lumber boards is not uniform in all dimensions. Wood shrinks most in the direction of the annual growth rings (tangentially), about half as much across the rings (radially), and only slightly along the grain (longitudinally). The combined effects of radial and tangential shrinkage can distort the shape of wood pieces because of the difference in shrinkage and the curvature of annual rings.

The reduction in size parallel to the growth ring, or circumferentially, is called tangential shrinkage, while the reduction in size parallel to the wood rays, or radially, is called radial shrinkage (Figure 4.16) (Glass and Zelinka 2010). At the same moisture content, the shrinkage is higher in the tangential direction and lower in the longitudinal direction.

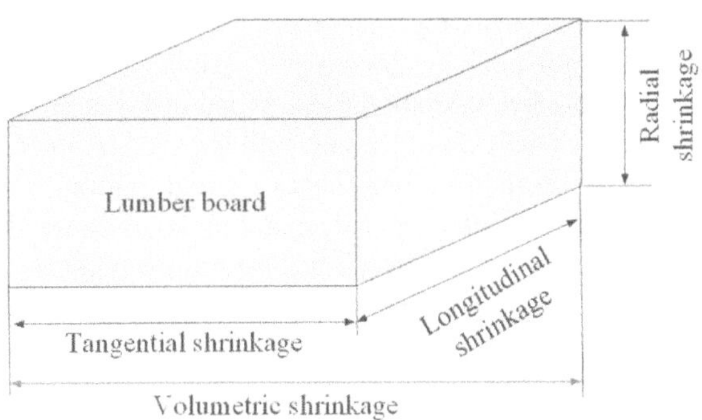

FIGURE 4.16 Tangential, longitudinal, and radial shrinkages of dried lumber boards.

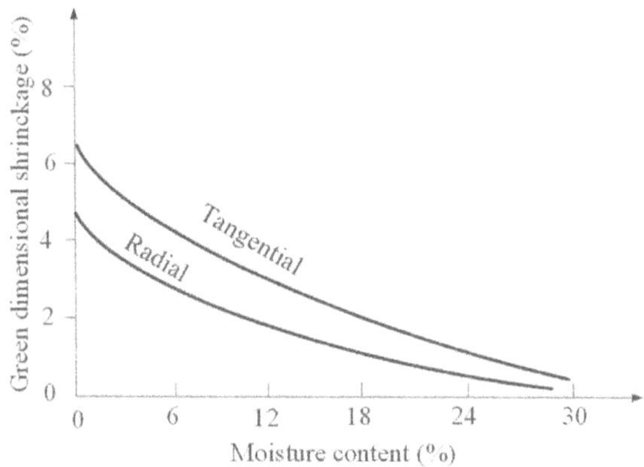

FIGURE 4.17 Typical wood shrinkage in tangential and radial directions. (Redrawn from Glass and Zelinka 2010.)

The effect is greater in the tangential direction (parallel to the growth rings) than in the radial direction (perpendicular to the growth rings). Dimensional changes in the longitudinal direction (parallel to the grain) are smaller and are usually insignificant in practice (Figure 4.17). A "rule-of-thumb" guide for the ratios of the dimensional changes in the three directions is: longitudinal/radial/tangential = 1/50/100 (Walker et al. 2013). The combined effects of radial and tangential shrinkage can distort the shape of wood pieces (Glass and Zelinka 2010).

The amount of shrinkage tends to increase with wood basic density. The denser the wood, the greater the shrinkage expected for a given change in moisture content.

As wood dries, shrinkage continues until the EMC is reached.

Wood is dimensionally stable when moisture content is greater than the fiber saturation point. Below MC_{fs}, wood changes dimension as there is loss of moisture (shrinks), because volume of the cell wall depends on the amount of bound water.

When wood is above fiber saturation point, the cell wall is in a nonshrunken condition and wood is capable of holding a large quantity of water.

Below fiber saturation point, as water is removed from the cell wall by diffusion, the micro-fibrils are drawn closer together, and the cells shrink. Volumetric change in wood below fiber saturation point is directly proportional to the volume of water lost from the cell wall. Below fiber saturation point, wood changes dimension as it loses moisture (shrinks), because volume of the cell wall depends on the amount of bound water. This shrinking can result in warping, checking, and splitting of wood. Below the fiber saturation point, wood shrinks with a decrease in the moisture content.

Volumetric shrinkage of wood below fiber saturation point is directly proportional to the volume of moisture (water) lost from the cell walls and can be expressed as follows (Cech and Pfaff 2000):

$$S_v = SG \cdot \left(FSP - MC \right) \quad (4.10)$$

where

S_v is the total specific volumetric shrinkage ($m^3/kg_{oven-dry}$);
SG is the basic specific density of wood ($m^3/kg_{green\,wood}$);
FSP is the fiber saturation point ($kg_{water}/kg_{oven-dry}$); and
MC is the moisture content of wood below fiber saturation point ($kg_{water}/kg_{oven-dry}$).

4.8.2 Other Wood Drying Defects

In controlling the wood drying defects and quality, the following factors and/or parameters must be carefully considered (Derome et al. 2015; Perré and Keey 2015): (i) adequate control of drying rates from the moment the lumber is sawn until the moisture content decreases is below 40% (d.b.); (ii) properly choose the drying schedules; (iii) surface moisture vaporization should be retarded through the employment of high air relative humidity while maintaining the highest safe rate of internal moisture movement by supplying more heat; (iv) carefully select the maximum drying temperature to which a given wood species can be dried (depending on the material thermal sensitivity); high values of dry-bulb temperatures may accelerate the internal moisture movement which could increase the risk of collapse, discoloration, or, even, thermal degradation of the wood; (v) provide low moisture content gradients by reducing the wet-bulb depression as much as possible; wood drying schedules generally must never exceed dry-bulb temperatures and temperature depression of 115°C and 45°C, respectively; (vi) provide high enough air velocity for good uniformity of drying throughout the lumber stack(s), even it may increase the electricity consumption and produce excessively high external heat transfer rates; and (vii) if stresses are severe enough that checks occur on the wood surface, rewetting the surface with wet steam (with hardwood) or water (with softwoods) may help the lumber surface to swell slightly, relieving the stress.

Differential drying between the lumber board surface and the core combined with heterogeneous shrinkage can result in additional defects: (i) checks, which can develop into serious degrade if the lumber is cycled between humid and dry conditions, such as caused by long fan reversal cycles at air relative high humidity; (ii) surface checks, such as results of using too low air relative humidity and/or too high air velocity; (iii) end checks, because the rapid longitudinal movement of moisture causes the board end to dry very quickly, developing high stresses; (iv) collapses, common in low-density woods with high green moisture content, which consist of distortion, flattening, or crushing of wood cells; the water is pulled out of the cells so fast that the cells collapse inward, reducing the lumen size. Very slow drying is the only potential preventive measure of collapse; however, almost all collapse can be recovered at the end of drying by steaming the lumber at 100% relative humidity; (v) warps (defined as deviations of the lumber boards' faces or edges from flatness, or any edge that is not at right angles to the adjacent face or edge), generally the result of drying at an excessively high relative humidity; and (vi) color change (discoloration, darkening), as a result of oxidative and enzymatic reaction with chemical constituents.

Almost all defects of dried wood can be prevented by proper design and operation of dryers.

With advances in research and development in wood drying, the fundamentals of the drying process and mechanisms of drying defects occurrence are much better understood, which helps to improve drying operations and develop new drying technologies.

REFERENCES

Anderson, J.O. 2014. Energy and resource efficiency in convective drying systems in the process industry. Doctoral thesis, Lulea University of Technology, Sweden.

Arends, T., J. Ahmed, L.P. Barakat. 2018. Moisture transport in pine wood during one-sided heating studied by NMR. *Experimental Thermal and Fluid Science* 99:259–271.

Beall, F.C. 1968. Specific heat of wood: Further research required to obtain meaningful data. U.S. Forest Service Research Note FPL-0184, February, USA.

Bian, Z. 2001. *Airflow and wood drying models for wood kilns.* A thesis submitted in partial fulfilment of the requirement for the Degree of Master of Applied Science in Faculty of Graduate Studies, Department of Mechanical Engineering, University of British Columbia, Canada.

Cech, M.J., F. Pfaff. 2000. *Operator Wood Drier Handbook for East of Canada*, edited by Forintek Corp., Canada's Eastern Forester Products Laboratory, Vancouver, Canada.

Crank, J. 1975. *The Mathematics of Diffusion*, Clarendon Press, Oxford, pp. 28–43.

Derome, D., A. Rafsanjani, S. Hering, M. Dressler, A. Patera, C. Lanvermann, M. Sedighi-Gilani, F.K. Wittel, P. Niemz, J. Carmeliet. 2015. The role of water in the behavior of wood. *Journal of Building Physics* 36(4):398–421.

Desch, H.E, J.M. Dinwoodie. 1996. *Structure, Properties, Conversion and Use.* Timber, 7th edition, Macmillan Education, London, UK.

Dinwoodie, J.M. 1981. *Timber, Its Nature and Behaviour.* Van Norstrand Reinhold, New York, NY.

Dunlap, F. 1912. The specific heat of wood. *Forest Service Bulletin 110*, U.S. Department of Agriculture, Washington, DC.

Eekelman, C.A. 1968. Wood moisture calculations. Purdue Forestry & Natural Resources, Department of Forestry & Natural Resources, FRN-156b, Purdue University, Indiana, USA (https://www.agriculture.purdue.edu/fnr/faculty/eckelman/documents/FNR-156b.pdf, accessed May 15, 2025).

Franke, B., S. Franke, M. Schiere, A. Müller. 2016. Moisture diffusivity in wood: Experimental and numerical investigations. *World Conference on Timber Engineering WCTE 2016*, Vienna, Austria, August 22–25.

Glass, S.V., S.L. Zelinka. 2010. *Moisture relations and physical properties of wood.* In: Wood *H*andbook: Wood as an Engineering Material, Centennial edition, General Technical Report FPL-GTR-190, United States Department of Agriculture, Forest Service, Forest Products Laboratory, Madison, WI, pp. 4.1–4.19.

Hajian, E., J.A.J. Huber, L. Hansson, D. Sandberg. 2024. High temperature drying of sawn timber: A review. *Drying Technology* 42(9):1397–1414.

Harley, J., J. Marchant. 1995. Methods of determining the moisture content of wood. Research Division, State Forests of New South Wales, Beecroft, Australia.

Hilderbrand, R. 1989. The drying of sawn timber, Maschinenbau, Niirtingen, Germany, 240 pp.

Keey, R.B., T.A.G. Langrish, J.C.F. Walker. 2000. Kiln Drying of Lumber, Springer Series in Wood Science, Springer, New York, NY.

Krabbenhoft, K. 2003. Moisture transport in wood: A study of physical-mathematical models and their numerical implementation. Ph.D. thesis, Department of Civil Engineering Technical University of Denmark, Copenhagen, Denmark.

Langrish, T.A.G., P.C.S. Kho, J.C.F. Walker. 1992. Experimental measurement and numerical simulation of local mass transfer coefficients in timber kilns. *Drying Technology* 10:753–781.

Minea, V. 2004. Heat pumps for wood drying: New developments and preliminary results. In: *Proceedings of the 14th International Drying Symposium*, Volume B, pp. 892–899, Sao Paulo, Brazil, August 22–25.

Minea, V. 2005. Laboratory and industrial-scale high temperature heat pump prototypes for wood drying. In: *Proceedings of Inter-American Drying Conference (IADC 2005)*, Montréal, Canada, August 21–23.

Minea, V. 2019. *Industrial Heat Pump-Assisted Wood Drying*. CRC Press, Boca Raton, FL.

Minea, V., D. Normand. 2005. Séchage de bois avec pompe à chaleur basse température (in French). Hydro-Québec & FP Innovation R&D report.

Mujumdar, A.S. 1995. *Handbook of Industrial Drying*, 2nd edition, Marcel Dekker, Inc., New York.

Mujumdar, A.S. 2015. Principles, classification, and selection of dryers. In: *Handbook of Industrial Drying*, 4th edition, edited by A.S. Mujumdar, CRC Press, Boca Raton, FL.

Perré, P., R. Keey. 2015. Drying of wood: Principles and practice. In: *Handbook of Industrial Drying*, 3rd Edition, edited by A.J. Mujumdar, Dekker, New York, NY.

Siau, J.F. 1984. Transport Processes in Wood, Springer Series in Wood Science, Springer, Berlin, Germany.

Skaar, C. 1988. *Wood–Water Relations*, Springer Publishing Series in Wood Science, edited by T.E. Timell, Springer, Jersey City, NJ.

Walker, J.C.F. 1993. *Primary Wood Processing Principles and Practice*, Chapman and Hall, London, 595 pp.

Walker, J.C.F., B.G. Butterfield, J.M. Harris, T.A.G. Langrish, J.M. Uprichard. 2013. *Primary Wood Processing: Principles and Practice*, Springer Science + Business Media, New York, NY.

5 Heat Pump-Assisted Drying

What's Wrong?

5.1 INTRODUCTION

Over the last few decades, heat pump-assisted drying systems were increasingly proposed and/or used in agro-food and wood processing industries owing to their lower energy consumptions, less quality losses of dried products, and higher drying performances compared to traditional drying methods. This energy-saving technology is based on the usage of (heat pump) reversed Carnot cycle principle that allows to recover energy from hot and humid drying air to supplement the energy required for drying (Mujumdar and Devahastin 2000; Mujumdar and Huang 2007; Chua et al. 2002; Chua et al. 2010; Colak and Hepbasli 2009).

Heat pump-assisted drying technique presents several advantages: (i) efficient control of drying air parameters (temperature and relative humidity); (ii) higher energy efficiency; (iii) higher drying performances; (iv) less material quality losses; and (v) not affected by outside weather as it uses practically closed drying chambers.

Combined with other drying methods (as solar, infrared, and vacuum), heat pump-assisted dryers may become more energetically efficient and more rapid compared to stand-alone heat pump-assisted dryers.

A brief review of published studies during the last few decades shows the missing details of proposed (developed) configurations and design/sizing aspects needed to accurately operate laboratory and industrial scale heat pump-assisted dryers. Many other drawbacks were also observed: relatively many published studies proposed complicated dryer–heat pump-coupled configurations; the quantities of dried materials have often been too small compared to heat pump nominal sizes; the reported drying performances were relatively high.

5.2 OUTLOOK ON DRAWBACKS OF SOME PAST PUBLISHED WORKS

In the past, many academic R&D studies on heat pump-assisted drying have been published that aimed to reduce the consumption of fossil energy, improve the overall energy efficiency of drying processes, and enhance the output quality of various thermo-sensitive dried products.

This chapter highlights the relevance and feasibility, as well as the degree of complexity and industrial applicability, of some proposed (and published) concepts. The aim is to better design (size), operate, and control heat pump-assisted drying systems in the future (Minea 2015a, 2015b).

 DOI: 10.1201/9781003618386-5

5.2.1 Incomplete Representation of Heat Pump-Assisted Drying Principle

As shown in Figure 5.1a (Chua et al. 2002), several published reviews and research studies on heat pump-assisted drying technology incompletely represented the basic principle of heat pump-assisted drying systems.

As can be seen in Figure 5.1a and b, the drying air leaving the heat pump's condenser enters the drying chamber at state d where it removes moisture from the dried material by an isenthalpic process $d-i$. Then, the humid air at state i is directed to the heat pump's evaporator where it is cooled (process $i-a$) and dehumidified (process $a-b$).

At the same time, the slightly superheated refrigerant enters the compressor at state 1 where it is adiabatically compressed (process $1-2$) to a higher pressure and temperature (state 2) (Figure 5.1c). The compressed superheated refrigerant vapor enters the heat pump's condenser that cools it up to the vapor saturation state $2a$ before being fully condensed (process $2a - 2b$) and subcooled (process $2b-3$) up to state 3. An expansion valve is used to reduce the pressure of subcooled liquid refrigerant at constant enthalpy (process $3-4$). After the abrupt reduction in pressure and temperature, the refrigerant enters the evaporator in a two-phase state (state 4) where it vaporizes at constant pressure (process $4-4a$) up to vapor saturated (state $4a$), and

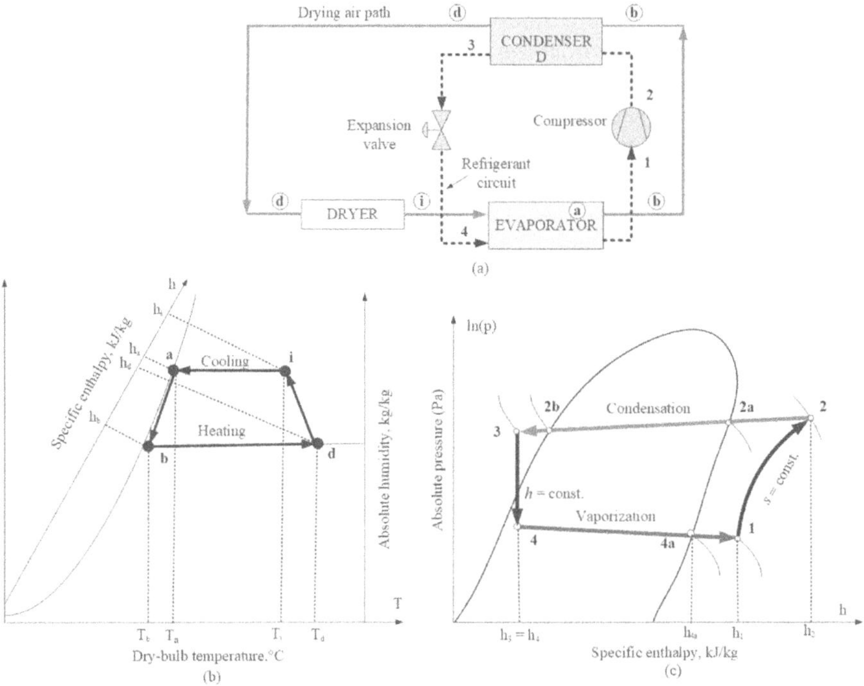

FIGURE 5.1 (a) Incomplete schematic representation of heat pump-assisted drying principle. (b) Inadequate representation of drying air thermodynamic processes in Mollier diagram. (c) Correct thermodynamic processes of refrigerant represented in ln(p)-h diagram. s, specific entropy.

then slightly superheated (process $4a-1$). This thermodynamic cycle will repeat until the drying process stops.

During the entire process 2–3 (Figure 5.1c), the heat released by the refrigerant condenser is used to heat up the drying air from state b to state d. Subsequently, the hot and dehumidified air at state d enters the drying chamber for drying purposes, and the drying air cycle repeats (Figure 5.1b).

The most significant drawback of schematics shown in Figure 5.1a and b is related to the drying air ducting across the heat pump evaporator and condenser. It can be seen that the drying air circulates at the same flow rate through the evaporator and the condenser. This is not a correct configuration because the drying air flow over the condenser has to be constant and higher than that flowing over the evaporator, while the drying air flow rate circulating over the evaporator must be varied during the entire drying cycle. If the drying air flow rate over the heat pump evaporator is too low, the heat pump compressor may trip off on "low" suction pressure control. On the contrary, if the airflow rate through the condenser is too low, the heat pump may trip off on "high" pressure control (Minea 2008).

5.2.2 Inappropriate Rejection of Excess Heat

Heat pump-assisted drying systems should be, as much as possible, physically isolated from the ambient environment to take full advantages of the technology, such as energy savings and environment protection. Heat pump-assisted drying systems that reject the excess heat from the refrigerant closed thermodynamic loop itself cannot be energetically efficient. The recovered heat should be added only to drying chambers to reach the set dry-bulb temperatures, and/or compensate the enclosures' heat losses.

On the other hand, in the case of occasional overheating of drying chambers, both inlet and outlet air vents must be automatically open. Also, when the dryer's fans shut down for few minutes during periodical change of their sense of rotation, the inlet and outlet air vents must also open. This operation aims to avoid the inward collapse (or implosion) of drying enclosures due to the sudden drops of the interior air pressure if the air vents remain closed. By neglecting such inevitable but relatively small heat losses through the air vents openings, heat pump-assisted dryers can be considered quasi-closed systems.

In several published studies, various methods were proposed and sometimes even used to expel the excess heat accumulated inside the heat pump thermodynamic cycles (Figure 5.2).

Because of the deficient control of drying processes (e.g., no drying air bypassing the evaporators), the heat accumulated (generated) inside the heat pump thermodynamic cycle is not completely absorbed by the dried materials at the same rhythm. As such, the excess amounts of heat are periodically rejected outdoors by using, for example, parallel condensers (Figure 5.2a) (Alves-Filho and Tokle 1999) or refrigerant desuperheaters (Figure 5.2b) (Alves-Filho and Eikevik 2008). It can be seen in Figure 5.2a that a three-way valve acts as a refrigerant modulating valve to direct the refrigerant flow to the heat pump internal condenser and then to the drying chamber, or to the external condenser to discharge the excess heat outside. When

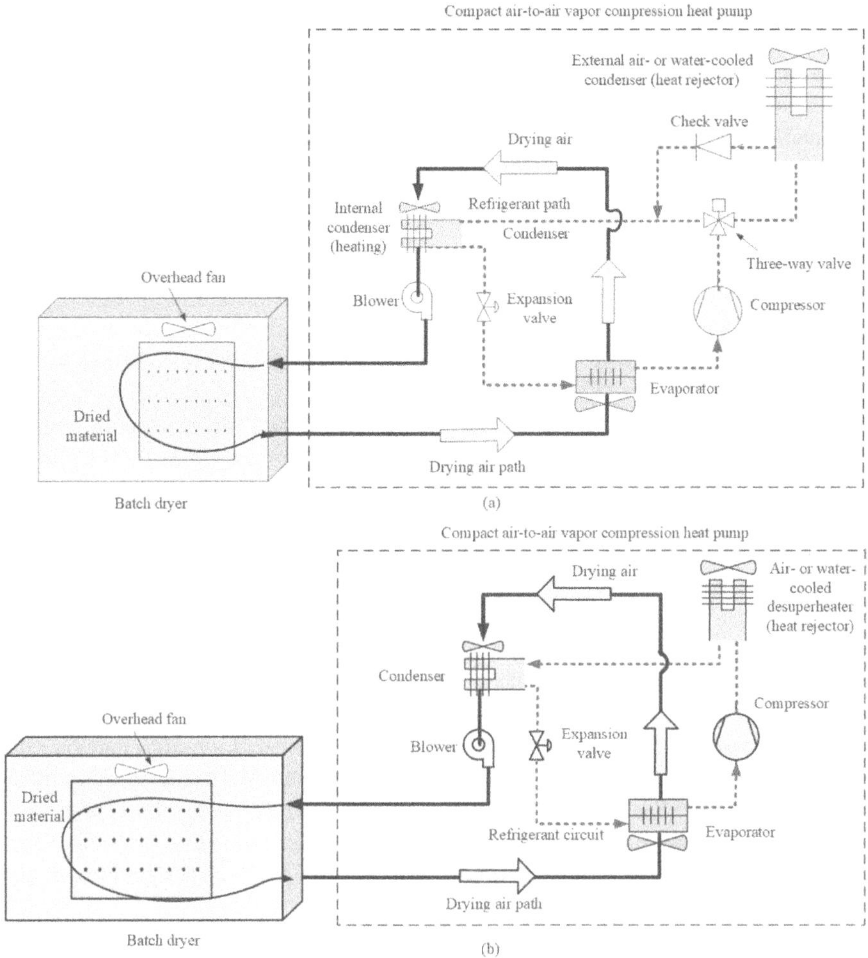

FIGURE 5.2 Published configurations of heat pump-assisted driers with heat rejection from heat pump thermodynamic cycles. (a) Heat rejection with external air- or water-cooled condenser. (b) Heat rejection with desuperheater.

the external condenser is operating, its blower as well as the heat pump compressor are also in operation, thus consuming electrical power despite no drying process occurring. Heat rejection with refrigerant air- or water-cooled desuperheaters (Figure 5.2b) could be a better solution but, depending on the cooling agent (air or water) temperature, it may reject sensible heat (15–20% of the total) and, possibly, some latent heat. In the latter case, the heating capacity of heat pumps is lowered because the refrigerant desuperheater and the internal condenser operate in series.

In conclusion, rejecting outdoors the excess heat from the refrigerant closed thermodynamic cycle with additional air- or water-cooled condensers, or refrigerant desuperheaters adulterates the role of heat pumps as drying devices and falsify the energy performances of heat pump-assisted drying systems. This means that heat pumps' coefficients of performance and dehumidification performances of heat pump-assisted drying systems cannot be accurately determined if (Minea 2007) (i) evaporator heat extraction capacity is too low compared to the material dewatering rate; and/or (ii) heat required for drying is mostly supplied by heat pump compressor.

5.2.3 Relevant Published Studies on Heat Pump-Assisted Drying Systems

Several published works on heat pump-assisted drying systems used configurations that made it possible to efficiently control the heat pump dehumidification capacities versus the products' thermal dewatering rates without rejecting the excess heat eventually accumulated inside the heat pump thermodynamic cycle (Minea 2010a; Minea 2010b).

One of the most relevant publications suggested to use for heat pump-assisted drying systems configurations as (Saensabai and Prasertsan 2003) (i) fully open; (ii) partially open; and (iii) fully closed.

The authors reasonably noted that (i) "the system configuration must be changed according to the changing properties of the working fluid (air) that are dependent on the product load, typical drying rate, and the ambient thermal conditions"; (ii) "there is no specific configuration for any single product, but the operation must be changed with respect to the product drying rate and ambient conditions"; (iii) "no general rule for optimum operation (minimum energy consumption or maximum drying rate) can be formulated"; (iv) "evaporator air bypass provides an opportunity for different heat recovery capacity and moisture content of the working air"; and (v) "the performance of heat pump-assisted drying systems are not well understood in many published works". The authors also highlighted experimental results of a closed-loop heat pump-assisted dryer with air bypassing the heat pump evaporator by using (i) 2-row, 2-circuit coils; (ii) refrigerant mass flow rates of 16–20 g/s per circuit; and (iii) drying air flow rates of 0.6–0.8 kg/s per circuit.

A compact heat pump-assisted batch dryer operating under constant drying rate conditions was modelled by Pal and Khan (2008a). The schematic drawing is relatively simple, and the impact of evaporator air bypass ratios is discussed. Relevant data such as initial material quantities, initial and final moisture contents, air drying temperatures and relative humidity, drying times, heat pump coefficients of performance ($COP_{heat\ pump}$), and system drying performances ($SMER$) are provided. The heat pump dehumidification rates, as well as the drying air temperatures and relative humidity are controlled by bypassing the heat pump evaporator.

Ong and Law (2009) noted that, in most cases, intermittent heat pump-assisted drying of fruits and vegetables improves the final quality of dried products. Chua (2000) reported that using intermittent heat pump-assisted drying concepts for drying agricultural products may shorten drying times by up to 61% as compared to a constant temperature drying schedules. Intermittent drying, based on the compressor hourly percentage running profile according to the drying air set and actual wet-bulb temperatures, was used for wood drying (Minea 2011; Minea 2012; Minea 2019). In this case, the heat pump's compressor and blower ran only when the actual (measured) drying air wet-bulb temperatures is within higher and lower preset limits.

According to such preset values, the compressor running time can be increased when the actual wet-bulb temperature is close to or higher than the upper limit, and decreased when it is close to or below the lower limit. The heat pump electrical energy consumption (compressor and blower) varied between 25% and 30% of the system total energy consumption. The dryer fan assumed 8–9% and the electrical or fossil backup energy assumed between 62% and 66% of the system total drying energy consumptions.

5.2.4 Complicated, Unproved, and/or Unreliable Heat Pump-Assisted Drying Concepts

Relatively complicated proposed/reported/published configurations involved, among others, the following configurations (Minea 2013a): (i) two-stage series evaporator heat pump-assisted drying system coupled with a single drying chamber; (ii) two-stage modular heat pump-assisted dryer; (iii) two-cycle heat pump-assisted drying system; (iv) two-stage drying system with series transcritical CO_2 heat pumps coupled with a single drying chamber; (v) two-stage fluidized bed heat pump-assisted dryer; (vi) ground-source heat pump-assisted drying system; and (vii) thermosyphon (heat pipe) heat pump-assisted drying system.

Based on Perry's concept (Perry 1981), Chua et al. (2002) developed and tested a drying system defined as a "two-stage series evaporator heat pump system for drying *with a single drying enclosure*"' (Figure 5.3).

This system, which contains two series (high- and low-pressure) evaporators (the high-pressure evaporator being intended to recover most of sensible heat from the drying air up to its dew point and the low-pressure evaporator to remove the latent heat), features a high degree of complexity and, therefore, few chances for industrial applications: First, because it uses two series evaporators with the same drying air flow rate coming from a single drying chamber. Second, because by splitting the heat pump evaporator in two (series) parts, the control of the heat pump dehumidification

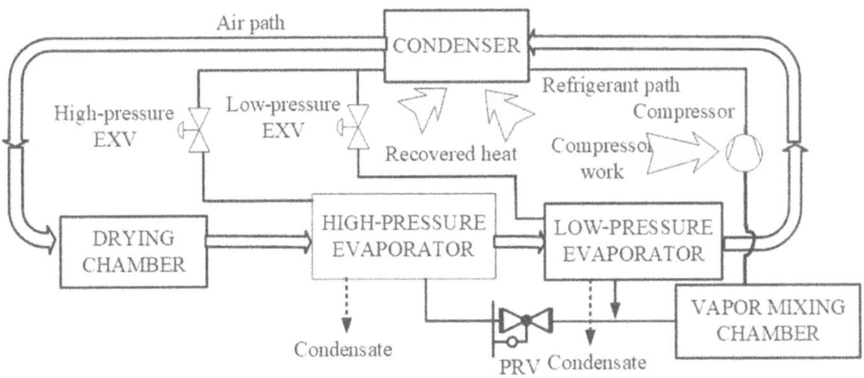

FIGURE 5.3 Two-stage series evaporator heat pump-assisted system coupled with a single drying chamber. *EXV*, expansion valve; *PRV*, pressure regulating valve. (Reprinted from Chua et al. 2002.)

capacity becomes practically impossible since the hot and humid air coming from the drying chamber does not bypass any of the two evaporators. Thus, no means are provided to vary the drying air flow rate aiming to control the system drying (dehumidification) capacity according to the material actual dewatering rate. Third, any heat pump-assisted drying system can't work with the same drying air flow rate circulating over the evaporator(s) and condenser(s). Despite these drawbacks, the authors wrongly concluded that "with constant drying temperature of 35°C, an appreciable improvement of 155% can be realized at the heat recovery level." They attributed this improvement to the fact that "with lower drying air temperature, the ability of the evaporators to extract latent energy from the moisture-laden air is enhanced since the sensible cooling duty to reduce the air temperature to the dew-point temperature has been reduced."

A transcritical CO_2 heat pump-assisted dryer for achieving high-quality dried products was proposed by Alves-Filho and Tokle (1999). The schematic diagrams of this system show that the CO_2 gas cooler and evaporator operate with the same drying air flow rates. However, it is well-known that in CO_2 transcritical heat pumps, the drying air flow rates over gas coolers should be approximately ten times higher than that circulating over the heat pumps' evaporators. Even though this concept looks simple, it has relatively few chances for industrial applications since (i) if the dryer starts operating at −30°C, as claimed by the authors, moisture freezing on the evaporator finned heat transfer surface may become a significant issue without providing means for periodical defrost; and (ii) control of excess heat rejection and of system dehumidification capacity (evaporator thermal power) are not addressed.

A CO_2 heat pump-assisted drying system used to convert raw and cold-extruded cranberry and potato–turnip mixtures into high-quality instant products was investigated by Alves-Filho (2002). The author did not indicate whether the CO_2 heat pump operated under subcritical or transcritical cycle, and if internal and external condensers (or gas coolers) were used. Also, the author did not indicate either at what temperatures the drying air is reheated inside the condensers (or gas coolers), and whether the condensers (or gas coolers) are able to reheat the drying air to the desired temperature while the same drying air circulates through both evaporators and condensers (or gas coolers).

Klöcker et al. (2001) studied the feasibility of laundry heat pump-assisted dryers using CO_2 as a working fluid. They found that the CO_2 transcritical cycle is "suitable" for heat pump-assisted dryers due to the good environmental characteristics and thermal properties of CO_2. The authors compared the energy performance of two optimized transcritical CO_2 heat pumps with those of a conventional electrically heated dryer. They concluded, without presenting reliable experimental tests, that if drying time is not a key requirement, energy savings of about 65% can be achieved with transcritical CO_2 heat pump-assisted dryers. Also, if a higher water extraction rate is sought, energy savings of about 53% can be achieved, including fan energy consumption, but with much shorter drying times than those of the conventional system.

A laboratory laundry dryer using CO_2 as a working fluid was built and tested by Klöcker et al. (2002). The degree of complexity of the proposed transcritical CO_2 heat pump-assisted drying system is relatively low, but chances of widespread applications seem much lower. That is mainly because the drying air flow rate through the evaporator and the gas cooler is the same, and the gas cooler heat transfer surface (118 m^2) seems oversized compared to the evaporator heat transfer surface (30.1 m^2), probably to compensate for the lower drying air flow rate over the gas cooler finned heat transfer surface. In addition, no means of controlling the heat pump dehumidification capacity (evaporator thermal power) as well as the temperature and relative humidity of the drying air supplied to the rotary drum are provided while the heat pump excess heat is rejected outside. The experiments showed that CO_2 heat pump-assisted dryers used for drying laundry at 50–60°C exhibit energy savings potential between 53% and 65% compared to conventional high-temperature (130°C) drying methods. However, the experimental drying cycles with transcritical CO_2 heat pumps were relatively short (about 54 minutes), while up to 35% of each drying cycle worked in a transient regime. Despite this operating condition, the authors reported relatively high system-specific moisture extraction ratios (*SMERs*) between 1.54 kg_{water}/kWh and 2.05 kg_{water}/kWh.

Based on the experimental results reported by Klöcker et al. (2002), Sarkar et al. (2006) simulated a transcritical CO_2 heat pump-assisted laundry dryer. The authors reported that the dryer efficiency, the ambient temperature, the air mass flow rate, and the recirculation air ratio have significant impacts on system performance. The maximum *SMER* was obtained with evaporator bypass air ratio of 0.3. The authors concluded that by increasing the bypass air ratio, dehumidification rate can be increased.

Honma et al. (2008) presented a compact heat pump-assisted dryer using CO_2 as a refrigerant for domestic clothes washers/dryers. The schematic diagram of their prototype did not show means for dehumidification capacity (evaporator thermal power) control, while the gas cooler had a heat transfer surface about 40% larger than that of the evaporator for the same drying air flow rate. The heat pump instantaneous coefficient of performance ($COP_{heat\ pump}$) (defined as the gas cooler heating capacity divided by the compressor input power) was estimated at 4.07. According to the authors, the prototype is able to reduce the electrical energy consumption by 59.2% and the drying time by 52.5% as compared with an electrical heater drying system. Furthermore, the authors estimate that the drying time can be further reduced by 3% by keeping the refrigerant superheating temperature at 6–10°C.

A complex concept proposed by Li et al. (2009), apparently with few chances for industrial applications, was a two-stage drying system with (series) transcritical CO_2 heat pumps with two series evaporators controlled by one expansion valve (Figure 5.4).

The question is whether the low- and high-side refrigerant loops could be efficiently charged with CO_2 if they operate with different drying chambers and, probably, different products. The authors did not indicate how the system dehumidification capacity (evaporators' thermal powers) of each evaporator and the final temperature and relative humidity of each drying air stream are controlled. This drying concept, including two series CO_2 compressors, each followed by a gas cooler–evaporator tandem serving two different drying chambers, seems to be less pertinent technically because (i) in transcritical CO_2 heat pumps, the gas coolers generally require much higher airflow rates than the evaporators; and (ii) the two series CO_2 compressors may produce final discharge pressures higher than 13 MPa depending on the actual

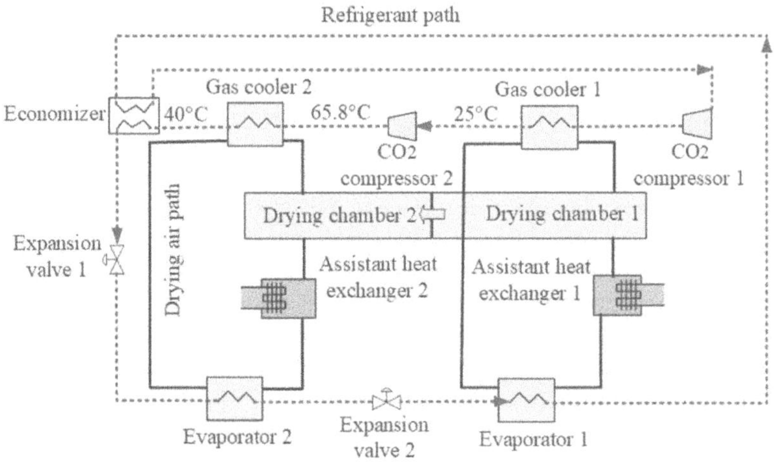

FIGURE 5.4 Diagram of a two-stage drying system with (series) transcritical CO_2 heat pumps. (Redrawn from Li et al. 2009.)

drying temperatures of each drying chamber and on the CO_2 compressors' compression ratios.

According to Fornasieri et al. (2010) and Mancini et al. (2011), transcritical CO_2 cycles are "viable alternatives to the traditional technology in the heat pump-assisted dryers" in terms of energy efficiency compared to subcritical heat pump-assisted dryers using HFC-134a as refrigerant. The system configuration is relatively simple but shows that the drying air flow rate through the evaporator and the gas cooler is the same. If so, the heat transfer area of the gas cooler must be much higher than that of the evaporator, leading to higher capital costs and unit cluttering. With such a configuration, the heat pump dehumidification rate (evaporator thermal power) and the temperature and relative humidity of the drying air supplied to the tumbler drum could not be efficiently controlled. The applicability of this concept seems relatively low because preheating steps may be required with long transient (stabilization) regimes up to 30% of the total drying cycles (i.e., 50–60 minutes each). Finally, the authors did not discuss the risk that the evaporator and gas cooler could get clogged with clothes fibers, dust, and other materials, nor the associated cleaning and excess heat rejection requirements.

Alves-Filho and Eikevik (2008) reported the results of drying porous materials as well as vegetables (green peas) and herbs in a laboratory hybrid fluidized bed heat pump-assisted atmospheric freeze-dryer. The concept has a low degree of complexity and relatively good chances of industrial applications. However, the heat pump fluidized bed dryer uses a desuperheater to reject the excess heat outside, but no estimate was provided about the actual amount of heat rejected and/or the additional energy consumed by the desuperheater fan(s). Also, the authors did not explain how the heat pump dehumidification capacity, the drying air temperature, and the relative humidity were controlled. The study did not provide information about the quantity of dried material and initial moisture content, heat pump nominal capacity, and drying schedules.

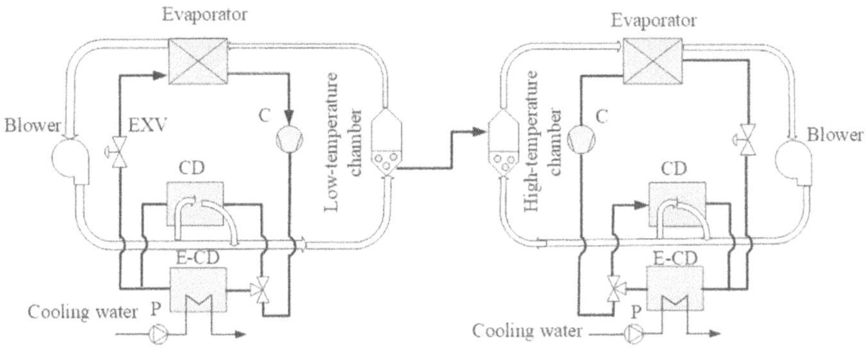

FIGURE 5.5 Schematic of a two-stage fluidized bed heat pump-assisted dryer. C, compressor; CD, condenser; E-CD, external heat rejection condenser; EXV, expansion valve; P, pump. (Redrawn from Alves-Filho and Tokle 1999.)

A relatively complex and costly two-stage fluidized bed heat pump dryer with two independent drying loops and two refrigerant (ammonia) circuits was reported by Alves-Filho and Tokle (1999) (Figure 5.5). Even though, according to the authors, the concept seems to have been experimented on a real-scale basis, chances of industrial applications seem limited because of some operational issues: (i) "low-temperature drying cycles are achieved below the product freezing point (i.e., –10°C)"; if so, the removed moisture is probably frozen on the finned surface of the first-stage evaporator; (ii) both external condensers are cooled with municipal costly freshwater; (iii) dried material "enters the second drying stage at its critical moisture content" without giving the amount of "critical moisture content"'; (iv) any indication is provided about how the high drying temperature of the second stage is attained, and, then, controlled; (v) any indication provided about how the drying (dehumidification) efficiency was affected by the product low initial moisture content; and (vi) any indication is provided about how the heat pump dehumidification capacity has been "varied."

Apparently simple but totally inefficient and unfeasible concept consists in using nonreversible ground-source heat pumps for drying apples (Colak and Hepbasli 2005; Kuzgunkaya and Hepbasli 2007a; Kuzgunkaya and Hepbasli 2007b) (Figure 5.6).

The main breakdowns of such a concept can be mentioned: (i) ground-source heat pump operates only in heating mode; (ii) no moisture is provided to the drying air inside the drying chamber; without moisture vaporization and dehumidification inside the drying chamber, no heat recovery from the drying air is achieved; (iii) for relatively small quantities of dried materials, this ground-source heat pump-assisted system could operate during relatively short drying cycles only in order to supply heat to the drying chamber at temperatures between 35°C and 45°C; however, in this range of drying temperatures, air-to-air mechanical vapor compression heat pumps may be technically and economically more competitive; (iv) for large-scale industrial systems with longer drying cycles, the concept, as proposed, is practically inapplicable without supplying auxiliary heat to the drying chamber; (v) in continuous (or long) drying processes, the ground temperature and thermal potential

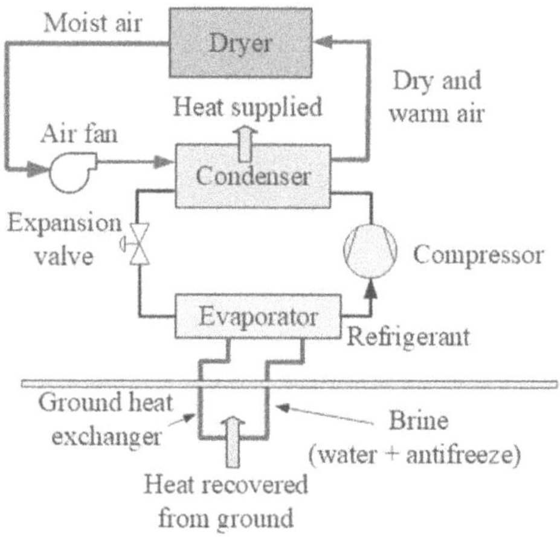

FIGURE 5.6 Schematic of a ground-source heat pump-assisted drying system. (Redrawn from Colak and Hepbasli 2005.)

will progressively decline; (vi) in this case, the heat pump will excessively cool and, then, freeze the ground around the vertical ground-coupled heat exchanger while the evaporating temperature will drop unduly, as well as the heat pump $\left(COP_{heat\ pump}\right)$; and (vii) after a certain period of time, the heat will be extracted from the ground at evaporating temperatures as low as −15°C, and even lower.

A complicated and costly heat recovery system was proposed by Phaphvangwittayakul et al. (2007) (Figure 5.7a). It combines an air-to-air mechanical vapor compression heat pump with a thermal loop (heat pipe). The experimental setup raises questions about how the evaporator and the heat pipe can be simultaneously and efficiently controlled to improve the system drying performances. As presented, the thermal loop may lead to a deterioration of the heat pump's overall coefficient of performance $\left(COP_{heat\ pump}\right)$ because the thermal loop simultaneously reduces the heat pump's evaporating temperature and increases its condensing temperature (Figure 5.7b). At the same time, such a configuration increases the compressor electrical input power and the system overall energy consumption.

Integration of air-to-air mechanical vapor compression heat pump in atmospheric freeze-drying fluidized beds was proposed as a new application concept (Alves-Filho et al. 2004; Eikevik et al. 2005; Strømmen and Kramer 1994). It is well-known that atmospheric freeze-drying is a convective drying process, occurring below the product freezing point at temperatures ranging typically from −3°C to −10°C. In spite of its apparent simplicity, the proposed dryer presents few chances of being implemented on industrial scales. The technical issues are as follows: (i) the proposed system does not make it possible to efficiently control the drying air supply temperature at the condenser outlet nor the heat pump dehumidification rates (evaporator thermal powers), mainly because the drying air is entirely recirculated through heat pump and that no air (or heat) is discharged from the drying chamber; and (ii) as represented, the experimental setup seems providing unbalanced airflows through both heat pump evaporator and condenser.

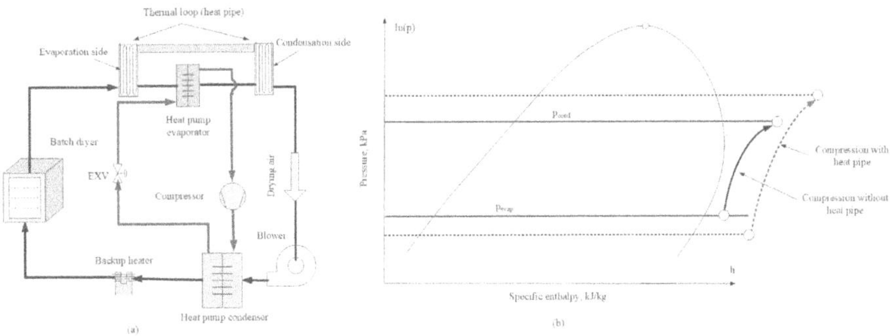

FIGURE 5.7 Schematic of a heat pump-assisted drying system with thermosyphon (heat pipe). (Redrawn from Phaphvangwittayakul et al. 2007.)

5.2.5 Data Provided Versus Results Reported

Efficient use of mechanical vapor compression heat pumps for drying depends on their optimum design according to the materials' moisture kinetics and drying rate curve(s). Pertinent theoretical and/or experimental results may not be provided without appropriate correlation between the dried material mass and moisture content, and the heat pump dehumidification capacity (evaporator thermal power).

This section reviews the input data provided (e.g., product initial and final mass, initial and final moisture contents, heat pump capacity), which is required to reasonably validate the relevance of reported results (e.g., drying time and dehumidification performance) in several theoretical laboratory and experimental published R&D studies on heat pump-assisted drying processes (Minea 2013b; Minea 2015b).

It was observed that several published works during the last few decades did not specify the heat pump–dryer coupling and quantities of dried materials while reporting high drying (energy) performances. Some of them include general statements such as "the drying conditions are controlled by adjusting the capacity of the heat pump components," without indicating "how" the capacity control was performed in particular cases.

On the other hand, information about initial warming-up (preheating) periods, as well as about the values of wet-bulb temperature of drying air during drying cycles, is not provided. In addition, information about how the material moisture vaporization rates were matched with the actual moisture condensation rates of heat pumps' evaporators was also not provided. Moreover, many published studies used means to reject outside the excess heat of heat pump thermodynamic closed loops; even this, in practice, probably seriously affected the reported overall drying (energy) performances.

5.2.5.1 Fruits

During the past few decades, many published journal articles and conference papers often reported "fabulous" energetic performances of fruit and vegetable heat pump-assisted (or combined/hybrid) drying systems (Salehi 2021), generally, without providing pertinent experimental and theoretical justifications.

For example, without valuable experimental proofs, it was stated that solar-assisted heat pump dryers increase the specific moisture extraction rates with lower specific energy consumptions (Şevik 2014). Used for drying banana slices, it was concluded that such systems provide "better" $COP_{heat\ pump}$ and higher *SMER* than the simple heat pump-assisted dryers. The average drying rates of simple heat pump-assisted dryers and solar-assisted heat pump dryers were found to be $0.205\ kg_{water}/kg_{dry\ solid} \cdot min$ and $0.342\ kg_{water}/kg_{dry\ solid} \cdot min$, respectively.

A (simulation) study (Kuan et al. 2019) reported that the heat pump-assisted dryers decrease the initial moisture content of banana from 74% (d.b.) to the final moisture content of 19% (d.b.) in 21 hours without specifying the type and nominal capacity of the mechanical vapor compression heat pump, as well as the mass of dried material. On the other hand, a conventional solar dryer reduced the initial moisture content from about 74% (d.b.) to the final moisture content of about 20% (d.b.) in 35 hours.

Aktaş et al. (2019) reported that using a heat pump-assisted dryer for drying grape pomace with drying air velocity of 2 m/s provided "better"' results than other applications in terms of bioactive properties of samples without explaining why (or providing experimental proofs). In addition, energy consumption of grape pomace dried by the heat pump-assisted dryer decreases by 51% compared with that of a hot air conventional convective dryer.

Zsivanovits et al. (2013) reported, without providing valuable information about the experimental conditions, that drying red grape samples at low temperatures (e.g., 45°C) is more suitable ("better") for product quality and bioactive properties than drying them at, for example, 50°C.

Liu et al. (2020) examined the effect of ultrasonic power on water removal kinetics and moisture migration of kiwifruit slices during ultrasound intensified the drying performances of heat pump-assisted dryer. The results showed that "contact ultrasound" application reduces the drying time significantly, and the increase of ultrasonic power has stronger reinforcing effect on dehydration rate.

About 800 g of grapes were dried with a "pilot-scale plant, comprising a closed-circuit, hot air convection chamber with heat pump" (Velazquez et al. 1997). Even though the heat pump refrigerant was identified (HFC-134a), the heat pump size or capacity (or at least the nominal thermal of the heat pump evaporator), as well as the moisture extraction rate (drying rate) (kg_{water}/h), the heat pump coefficient of performance ($COP_{heat\ pump}$), or the dehumidification performance of drying system ($SMER$), was not (Table 5.1). Therefore, the claimed reduction of the drying time from 40 hours (normally required with sun drying) to 24 hours performed with heat-pump-assisted drying system seem not credible.

For a pilot-scale heat pump-assisted dryer that aimed to dry 50 kg of peas was used by Rahman et al. (1998); data on air drying temperature and relative humidity ranges were provided, but not the material initial and final moisture contents, heat pump capacity (e.g., compressor nominal input power), and drying time (Table 5.1). On the other hand, despite the missing essential experimental data, dehumidification performances (*SMERs*) as high as 8 have been reported. Finally, the system start-up (warming-up) operation was not clearly explained and, consequently, the reader may be led to understand that the heat pump reached the operating parameters without any product inside the drying chamber.

TABLE 5.1
Data Provided versus Results Reported for Heat Pump-Assisted Fruit Drying

Dried Product (reference)	Data Provided: Product Initial Mass	Data Provided: Moisture Content ($kg_{water}/kg_{dry\ solid}$) Initial	Data Provided: Moisture Content ($kg_{water}/kg_{dry\ solid}$) Final	Data Provided: $\dot{W}_c$ ($\dot{Q}_{ev}$)	Results Reported: T (RH)	Results Reported: Moisture Extraction Rate	Results Reported: Drying Time	Results Reported: SMER (COP_{hp})
–	–	–	–	–	–	–	–	–
–	kg	%	%	kW (kW)	°C (%)	kg_{water}/h	hours	kg_{water}/kWh
Grapes Velazquez et al. (1997)	0.8	3.5	<0.5	No	Yes	No	20–80	No (No)
Pears Rahman et al. (1998)	50	No	No	No	25–65 (20–60)	1–4	No	1–8
Green peas Alves-Filho et al. (2004)	No	73 (w.b.)	No	No	–6; –3; 0; 25	No	12	No (No)
Green peas Zielinska et al. (2012a, b)	No	3.23	0.09	No	–5°C (20–55)	4.8	Yes	No (No)
Tropical fruits Ceylan et al. (2007)	No	1.5-4.7	0.3–0.7	No	40	No	6	No (No)
Nectarines Sunthonvit et al. (2007)	No	No	18–20 (w.b.)	No	25 (10)	No	19	No (No)
Figs Xanthopoulos et al. (2007)	0.4	71.3 (w.b.)	30 (w.b.)	2	Yes	No	23–53.3	No (No)
Plum Hepbasli et al. (2010)	No	84.5 (w.b.)	15.7 (w.b.)	No	45–55	No	No	No (2.81)

Note: COP_{hp}, heat pump coefficient of performance; T, drying temperature; *No*, information (data) not provided; $\dot{Q}_{evap}$, evaporator thermal capacity; RH, relative humidity; *SMER*, specific moisture extraction ratio; *w.b.*, wet basis; $\dot{W}_c$, compressor-rated input power; *Yes*, information (data) provided.

Green peas were dried "in a heat pump-assisted dryer under atmospheric freeze-drying conditions (–6°C, –3°C, and 0°C) and at 25°C, operating either at constant or combination of temperatures" (Alves-Filho et al. 2004). It can be seen that this paper did not provide information about the refrigerant used, the heat pump operating parameters (pressures, temperatures), initial quantities and final moisture content of the dried product, moisture extraction rates (kg_{water}/h), heat pump coefficients of performance ($COP_{heat\ pump}$), and the drying system dehumidification performances (*SMERs*) (Table 5.1). In addition, even some experiments have been performed at very low temperatures such as –6°C, 3°C, and 0°C. The authors reported that

> the drying air passed through the bed of particles and removed moisture from the wet material. Then, the moistened air flowed through the filter and air cooler (evaporator) where it was cooled below its dew point leading to condensation of water vapor. The evaporator absorbed latent heat of condensation of moisture to boil the fluid inside the tubes.

Or, it can be noted that when drying with heat pump-assisted systems at temperatures near or below approximately 12°C, the refrigerant evaporating temperatures are much lower than the drying air dew points corresponding to drying air operating temperatures at normal atmospheric pressure. Consequently, in such circumstances, the moisture removed from the dried material may freeze on the surface of the evaporator fins and, therefore, complex and costly defrosting procedures must be periodically initiated. The authors did not mention whether this phenomenon occurred during the 12-hour freeze-drying process and how they achieved the evaporator defrosting cycles, if any. In addition, an experimental system drying performance (*SMER*) or a heat pump coefficient of performance ($COP_{heat\ pump}$) has been reported.

The effect of multistage heat pump-assisted fluidized bed atmospheric freeze-drying and microwave vacuum drying on the drying kinetics, moisture diffusivities, microstructure, and physical parameters of green peas was evaluated by Zielinska et al. (2012a, 2012b). Even if the initial mass of dried green peas is not specified, the initial moisture content (about $3.23 \pm 0.02\ kg_{water}/kg_{dry\ solid}$, d.b.) and the final moisture content ($0.09 \pm 0.04\ kg_{water}/kg_{dry\ solid}$, d.b.) were indicated (Table 5.1). According to the authors, the samples were placed in a freezer at –20°C before drying tests were performed with a mechanical vapor compression heat pump at –5°C, which is below the freezing point of the frozen green peas, and the inlet air relative humidity was kept at both a low (20%) level and a high (55%) level. However, the authors did not explain how they kept these parameters constant during each drying cycle. No information was provided on the initial quantity of dried green peas and the heat pump's compressor-rated input power ($\dot{W}_{comp}$) or evaporator nominal dehumidification capacity ($\dot{Q}_{evap}$), nor the results such as the heat pump coefficient of performance ($COP_{heat\ pump}$) and the system drying performance (*SMER*). These authors reported that with the combined method (i.e., fluidized bed followed by heat pump-assisted drying), the drying rate was of about $4.8\ kg_{water}/h$. Even though many input data were missing, the authors reported results on product drying kinetics and the final morphological parameters, texture properties, and color. However, it would be impossible to validate such results without pertinent experimental data about the combined drying systems studied.

Ceylan et al. (2007) used a heat pump-assisted dryer for drying unknown quantities of tropical fruit (kiwi, avocado. and banana) at 40°C during 6 hours (Table 5.1). Even though the initial and final product moisture contents (expressed as $kg_{water}/kg_{dry\,solid}$) were provided, the system drying performance (*SMER*), heat pump size, e.g., compressor-rated input power ($\dot{W}_{comp}$) or evaporator nominal dehumidification capacity ($\dot{Q}_{evap}$), the heat pump coefficient of performance ($COP_{heat\,pump}$), as well as the moisture extraction rate or quantity of water removed were not mentioned.

According to Sunthonvit et al. (2007), the best systems for preserving the volatile compounds of dried nectarines are heat pump-assisted dryers. In this publication, the drying temperatures and the material final moisture content were provided, but not its initial mass and initial average moisture content (Table 5.1). Moreover, the heat pump compressor-rated input power ($\dot{W}_{comp}$) or evaporator nominal dehumidification capacity ($\dot{Q}_{evap}$), moisture extraction rate ($kg_{water}/kg_{dry\,solid}$), system drying performance (*SMER*), and heat pump coefficient of performance ($COP_{heat\,pump}$) were not provided. Therefore, the reader may not understand nor validate the author's conclusion stating that the heat pump-assisted dryer "followed by cabinet and tunnel drier" was the best system for "preservation of volatile compounds" of tropical fruits.

A small quantity of single-layer whole figs (400 g) was dried by Xanthopoulos et al. (2007) with a 2-kW (compressor-rated input power) heat pump using HCFC-22 as a refrigerant. The initial and final moisture contents (wet basis) of the dried product were provided as well as the experimental drying times (Table 5.1). Missing information includes drying schedules used (with or without preheating steps), moisture extraction rate ($kg_{water}/kg_{dry\,solid}$), as well as the system drying performance (*SMER*) and heat pump coefficient of performance ($COP_{heat\,pump}$).

Plum slices were dried at 45°C and 55°C in a heat pump-assisted conveyor dryer, but no experimental results were provided (Hepbasli et al. 2010). Even though the initial (84.5%, w.b.) and final (15.7%, w.b.) product moisture contents were provided, the initial mass of the dried plum slices, the heat pump size ($\dot{W}_{comp}$ or $\dot{Q}_{evap}$) and operating parameters, as well as drying times, and system dehumidification performances (*SMER*) are not provided (Table 5.1). In addition, the authors did not explain how the temperatures and relative humidity of drying air were kept constant. In spite of such essential missing data and results, the authors reported the heat pump coefficients of performance up to 2.8. In addition, a superfluous exergy analysis of the mechanical vapor compression heat pump cycle is performed without any apparent link with the heat pump-assisted system used to dry plums.

5.2.5.2 Vegetables

Many academic research studies have been published on drying vegetables with heat pump-assisted dryers, but several of them failed to provide all required data to support all or a part of the reported results.

For example, even though the initial mass (200 g) of the dried material (vegetable seeds) and the initial (30%, d.b.) and final (6%, d.b.) moisture contents were provided, the CO_2 heat pump's compressor nominal input power ($\dot{W}_{comp}$) or the heat pump's evaporator-rated thermal power ($\dot{Q}_{evap}$), the system dehumidification performance (*SMER*), and the heat pump coefficient of performance ($COP_{heat\ pump}$), as well as the moisture extraction rates and drying times, were not provided (Table 5.2) (Li et al. 2009). With an initial mass of dried material as small as 200 g, overall *SMERs* as high as 4.5 kg_{water}/kWh were obtained by (theoretically) varying the air mass flow rates and intermediate pressures. Such an unlikely performance, achieved with theoretical drying concepts involving both single- or two-stage series transcritical CO_2 heat pumps and two series drying chambers, presents little credibility for both academic R&D community and industrial drying industry.

Contrary to the previous study, Jinjiang and Wu (2007) provided a lot of useful information in their paper. The initial mass of the dried products (paddy and shredded radish) (1,200 kg) seems well-correlated with the compressor nominal input power (4.2 kW) and initial (30–35 $kg_{water}/kg_{dry\ solid}$) and final (18 $kg_{water}/kg_{dry\ solid}$) moisture contents (Table 5.2). The results reported, such as drying temperatures of 42°C and 46°C, drying times (15–16 hours), moisture extraction rates (8–15.9 kg_{water}/h), and average *SMERs* (2 kg_{water}/kWh), help to better validate the reliability of the experimental work, as well as the overall performance of the heat pump-assisted drying system.

The characteristics and product quality (especially the color change) of continuously and intermittently dried 1 kg of polypore mushroom (*Ganoderma tsugae*) from 203 $kg_{water}/kg_{dry\ solid}$ (w.b.) to 12 $kg_{water}/kg_{dry\ solid}$ (w.b.) at 28–40.6°C was investigated by Chin and Law (2010) (Table 5.2). First, no information was provided about the heat pump–rated capacity ($\dot{W}_{comp}$ or $\dot{Q}_{evap}$). Second, the authors did not indicate heat pump energy consumptions in both continuous and intermittent drying operating modes. Third, the study reported that the material was dried over 30–40 hours, while two graphical representations showed that the moisture ratios rapidly decreased by up to 70% of the initial values within the first 3–4 hours for all experimental intermittencies, whether the air heater was activated or not. Fourth, the heat pump operating parameters were not shown prior to analyze the final color of the dried product.

Queiroz et al. (2004) used mechanical vapor air-to-air heat pumps (with HCFC-22 as a refrigerant) to dry tomatoes at temperatures of 40°C, 45°C, and 50°C without providing information about the initial mass, initial and final moisture contents, and heat pump nominal capacity ($\dot{W}_{comp}$ or $\dot{Q}_{evap}$) (Table 5.2). On the other hand, the authors reported that "energy savings of about 40% with heat pump dryers were achieved," but no information on the heat pump operating parameters, drying schedules, moisture extraction rates, and drying times was provided. The system's dehumidification performances ($SMER = 2.56–2.68$) have been estimated, but the heat pumps' coefficients of performance ($COP_{heat\ pump}$) (based on compressor and drying air fan electrical powers) were not estimated.

Batch and continuous bed dryers with two parallel mechanical vapor compression heat pumps for drying crops (chopped alfalfa) were modeled and experimented by Adapa et al. (2002). The authors provided useful information about the heat pump

TABLE 5.2
Data Provided and Results Reported for Drying Vegetables with Heat Pump-Assisted Systems

	Data Provided				Results Reported			
Dried Product (reference)	**Product Initial Mass**	**Moisture Content ($kg_{water}/kg_{dry\ solid}$)**		**$\dot{W}_c$ ($\dot{Q}_{ev}$)**	***T* (*RH*)**	**Moisture Extraction Rate**	**Drying Time**	***SMER* (COP_{hp})**
–	–	**Initial**	**Final**	–	–	–	–	–
–	**kg**	**%**	**%**	**kW (kW)**	**°C (%)**	**kg_{water}/h**	**hours**	**kg_{water}/h**
Vegetable seeds Li et al. (2009)	0.2	30	6	No	30 (55) – 50 (50/80)	No	No	3.1–4.5 (No)
Paddy Jinjiang and Wu (2007)	1,200	30–35	13	4.2	42 and 46 (26 and 14)	8–15.9	15–16	2.0 (No)
Mushroom Chin and Law (2010)	1	203 (w.b.)	12 (w.b.)	No	40.6 and 28.4 (16.2)	Yes	30–40	No (No)
Tomato Queiroz et al. (2004)	No	No	No	No	40, 45, 50 (No)	No	No	2.56–2.68 (No)
Specialty crops Adapa et al. (2002)	11.6–20.7	63–89	5.1–10.8	No	30–35 (10–25)	Yes	Yes	0.06–0.61 (No)
Ginger Hawlader et al. (2006)	0.1	No	No	No	45 (10)	No	8	No (No)
Shredded radish Lee and Kim (2007)	200	95	16	15	40 (No)	6.3	25	1.5 (3–4)
Green sweet peppers Pal et al. (2008b)	0.025	145 (d.b.)	10 (d.b.)	No	30, 35, 40, 45 (19–55)	No	17–36	0.55–1.1 (No)
Saffron Mortezapour et al. (2012)	0.5	80 (w.b.)	10 (w.b.)	No (No)	40; 60	No	~0.5	0.5–1.1 (No)

Notes: COP_{hp}, heat pump coefficient of performance; d.b., dry basis; *T*, drying temperature; No, information (data) not provided; $\dot{Q}_{ev}$, evaporator thermal capacity; *RH*, relative humidity; *SMER*, system-specific moisture extraction ratio; *w.b.*, wet basis; $\dot{W}_c$, compressor-rated input power; *Yes*, information (data) provided.

refrigerant (HFC-134a), the heat pump (compressor)–rated input power ($\dot{W}_{comp}$), the evaporating and condensing pressures and temperatures (0.97/2.03 MPa and 40°C/68°C, respectively), initial product masses, initial and final moisture contents, drying times, maximum and minimum temperatures, and air relative humidity, as well as the mass of water extracted (Table 5.2). However, the experimental tests achieved poor *SMERs*, which varied between 0.5 kg_{water}/kWh and 1.02 kg_{water}/kWh. The authors explained this disappointing performance by the lesser drying rates of hygroscopic alfalfa compared to other nonagricultural products. Additional causes may include the fact that the heat pump evaporators did not have external fins and that the temperature inside the drying chamber was maintained at a constant level by exhausting part of the drying air from the cabinet, resulting in losses of latent and sensible heat. Obviously, the final conclusion stating that drying chopped alfalfa with heat pump-assisted dryers may decrease the moisture contents to levels as low as 10% (wet basis) seems inappropriate.

Ginger was dried in a modified atmosphere (air, nitrogen, carbon dioxide) with heat pump-assisted dryers and other (freezing and vacuum) drying methods (Hawlader et al. 2006). Despite their conclusions concerning the increase of "6-gingerol retention and flavour retention," the experimental method and the results obtained with a heat pump-assisted drying system seem less relevant. First, an initial mass as small as 100 g of ginger slices has been used in each of experimental drying process at 45°C. Even the drying cycles lasted 8 hours, but no information on the heat pump capacity (e.g., compressor nominal input power $\dot{W}_{comp}$ or evaporator dehumidification capacity $\dot{Q}_{evap}$), as well as the product initial and final moisture contents, the quantity of water removed, and drying performance ($COP_{heat\ pump}$, *SMER*) was provided (Table 5.2). In addition, the authors did not indicate whether changes in the heat transfer process at the heat pump evaporator and condenser levels were observed when normal air was replaced with nitrogen and carbon dioxide as drying mediums.

Contrary to other published studies, Lee and Kim (2007) provided more useful information about the dried product (shredded radish): (i) inlet mass (200 g); (ii) initial (95%) and final (16%) moisture contents: (iii) drying time (25 hours); (iv) moisture extraction rate (6.3 kg_{water}/h); (v) heat pump coefficient of performance ($COP_{heat\ pump} = 3-4$); and (vi) system dehumidification performance ($SMER = 1.5\ kg_{water}/h$) (Table 5.2). The authors found that the batch heat pump-assisted drying system took 1.2 – 1.5 times longer drying time than the conventional convective hot air dryers (i.e., 25 hours versus 15 hours). This result was attributed to the "insufficient capacity of the heat pump system used" versus the initial mass and moisture content of the dried material, even if the compressor nominal electrical input power (15 kW) seems excessively high (overdesigned) in comparison to the very low initial mass of dried material. Despite this aspect, the authors concluded that "the heat pump dryer showed considerable improvement in energy saving" because $SMER = 1.5\ kg_{water}/h$ was about three times higher than that of the hot air dryer using diesel fuel as a heat source.

Apparently, very small quantities (25 ± 2 g) of green sweet pepper thin layers were dried in a heat pump-assisted dryer at drying air temperatures ranging from 30°C to 45°C, and relative humidity from 19% to 55% during drying periods varying from 17 hours to 30 hours (Table 5.2) (Pal et al. 2008b). The most important missing data was the size of the heat pump (compressor-rated input power $\dot{W}_{comp}$ or evaporator dehumidification capacity $\dot{Q}_{evap}$). The tests reduced the material's moisture content up to 10% (d.b.) As a result, very low *SMERs* (0.55 kg_{water}/kWh at 30°C and 0.93 kg_{water}/kWh at 45°C) were achieved. Despite such poor dehumidification performances, the authors recommended to "dry green sweet peppers at 35°C with a heat pump dryer to obtain acceptable product quality attributes of dehydrated products."

A very small quantity (0.5 kg) of saffron stigma was dried with a "hybrid PVT solar dryer and a heat pump-assisted unit arranged to create flow air in a closed (or open) cycle current" (Mortezapour et al. 2012). In spite of some data which were provided about the heat pump-assisted hybrid dryers (e.g., initial [80%, w.b.] and final [10%, w.b.] moisture contents) (Table 5.2), this published work raises serious questions about its technical relevance. First, the ambient air entered the photovoltaic (PV) solar collectors at temperatures between 25°C and 38°C from 9 A.M. to 3 P.M., but the outlet temperatures were not indicated. After the photovoltaic solar collectors, the air is mixed "with the dryer air to drop its temperature and relative humidity" prior to entering the heat pump condenser. After the condenser, an auxiliary electrical heater reheated the air up to the desired set drying temperature. A logical question could be: why does the system have to cool down the air coming from the solar PV collectors (by mixing it with fresh air), and, then, heat it again with the heat pump condenser and/or electrical heater. Second, according to the authors, the heat pump dryer input power was entirely provided by the PV collectors (~120 W) during a relatively short single cycle per day (i.e., about 30 minutes). However, the drying system was intended to "dry the daily harvest from a farm in the area where the experiments were conducted" by using power from the grid for the additional drying cycle. Unfortunately, the authors didn't indicate how many additional cycles are required each day. And, if this number is relatively high, how could the economic efficiency of the proposed hybrid drying system, which provided maximum *SMERs* of 1.15 kg_{water}/kWh at 60°C during only one cycle per day be justified?

5.2.5.3 Woods

In countries such as the United States, New Zealand, Canada, and Poland, wood drying is one of the largest markets for heat pump-assisted drying systems. However, few experimental research studies were reported during the last few decades.

Carrington et al. (1995) analyzed the operating characteristics of a medium-temperature heat pump-assisted dryer using HFC-134a as a refrigerant with "passive evaporator-economized and staged liquid sub-cooling." The authors reported that the system maximum dehumidification performance (*SMER*) could reach 5.11 kg_{water}/kWh at 50°C dry-bulb temperature and 90% relative humidity (Table 5.3). However, this level of performance was provided without any indication about (i) material initial mass (or volume); (ii) material initial and final moisture contents; (iii) heat pump capacity ($\dot{W}_{comp}$) or $\dot{Q}_{evap}$); (iv) moisture extraction rates; and (v) drying times.

Chen et al. (2002) used a dual-purpose heat pump dehumidifier for drying 1,154 kg of high-value seed pine cones and pine pollen catkins. Both the low- and high-temperature kiln drying methods were explained and experimental results were provided based on the product initial mass and inlet/outlet moisture contents (Table 5.3).

TABLE 5.3
Data Provided and Results Reported for Drying Woods with Heat Pump-Assisted Batch Dryers

	Data Provided				Results Reported			
Dried Product (reference)	Initial Mass	Moisture Content ($kg_{water}/kg_{dry\ solid}$)		$\dot{W}_{comp}$ ($\dot{Q}_{evap}$)	T (RH)	Moisture Extraction Rate	Drying Time	SMER (COP_{hp})
–	–	Initial	Final		–	–	–	–
–	kg (m^3)	%	%	kW	°C (%)	kg_{water}/h	hours	kg_{water}/kWh
Woods Carrington et al. (1995)	No	No	No	No	50 (90)	No	No	5.11 (No)
Pine cones Chen et al. (2002)	1,154.4	40–45 (w.b.)	5–10 (w.b.)	No	25 (average13)	0.67 (average)	560 minutes	0.25 (No)
Softwood Minea (2004)	354 m^3	30–40 (d.b.)	~7 (d.b.)	130	82–85	300	60–151 hours	1.5–3.25 (3–4.6)
Hardwood Minea (2006)	8–13 m^3	31–76 (d.b.)	7.6 (d.b.)	5.6	37–43	4.3–9.2	Max. 206 hours	2.44–3 (No)

Notes: T, drying temperature; No, information (data) not provided; $\dot{Q}_{evap}$, evaporator normal thermal capacity; RH, relative humidity; *SMER*, specific moisture extraction ratio; w.b., wet basis; $\dot{W}_{comp}$, compressor-rated input power; Yes, information (data) provided.

The average moisture extraction rate varied around 0.67 kg/h only, i.e., approximately 1 kg/h lower than the predicted value. The authors explained this difference by stating that the evaporator frosted at relative humidity of drying air between 20% and 30%. To operate the dehumidifier under the required low-temperature conditions, the dehumidifier was provided with a reverse defrost cycle and additional valves to prevent possible compressor damages caused by the refrigerant liquid flowing backs. Finally, the cycle overall *SMER* was relatively low (0.25 kg_{water}/kWh) because of the oversized kiln fans used and (unknown, unpublished) the compressor nominal electrical power input.

Drying coniferous (resinous) lumber (e.g., white spruce, balsam fir, also known as softwood) with high-temperature heat pumps was reported by Minea (2011). A hybrid (electricity/fossil fuel) high-temperature heat pump-assisted drying system was developed and field tested. A conventional wood dryer made of insulated panels was converted into a heat pump-assisted dryer. This study demonstrated that the thermodynamic efficiency and energy performance of the high-temperature heat pump-assisted dryer reached the design targets. The average measured specific moisture extraction rate of the heat pumps was 2.35 kg_{water}/kWh (with white spruce) and 1.5 kg_{water}/kWh (with balsam fir), while the heat pump average coefficients of performance generally varied from 4.6 (at the beginning) up to 3.0 (toward the end of drying cycles). Cycle duration ranged from 2.5 days (with white spruce) to 6.3 days (with balsam fir), including the initial preheating steps (Table 5.3). Better insulated and well-maintained dryers are necessary to reach drying temperatures above 100°C as well as reduce the drying time by up to 25% and total energy consumption by up to 50% for resinous species.

In a published work about an intermittent heat pump-assisted system used for drying hardwoods (Minea 2004), the dryer maximum capacity, the initial volumes, and initial/final moistures contents of the wood, as well as the heat pump–rated capacity, the refrigerant used, the drying schedules, and many other useful information (e.g., water extraction rates, drying curves and energy performance) were provided. Deciduous trees (e.g., hard maple, yellow and white birch, oak, also known as hardwood) were dried with a low-temperature electrically driven mechanical vapor compression heat pump.

Intermittent drying, based on the compressor hourly percentage running profile according to the air set and actual wet-bulb temperatures, was used for wood drying (Minea 2006).

A 5.6-kW low-temperature heat pump coupled to a 13 m^3 hardwood dryer was tested with electricity and natural gas (steam) as backup energy sources. The heat pump (compressor and blower) electrical energy consumption varied between 25% and 30% of the total equivalent energy consumption of each drying cycle. The dryer central fan represented 8–9% and the electrical or fossil backup energy between 62% and 66% of the total equivalent drying energy consumption. For initial moisture contents above 41%, the total water quantities extracted above the fiber saturation point (25%) were up to 2.9 times higher than those removed below the fiber saturation point. Consequently, in these cases, the heat pump dehumidification efficiency was up to three times higher above the fiber saturation point compared to that obtained below the fiber saturation point. The hybrid drying cycles reduced natural gas consumption by 56% and the equivalent energy costs by 21.5%, compared to a conventional drying cycle with natural gas as unique heating energy source.

REFERENCES

Adapa, P.K., G.J. Schoenau, S. Sokhansanj. 2002. Performance study of a heat pump system for specialty crops – Part 2: Model verification. *International Journal of Energy Research* 26:1021–1033.

Aktaş, M., L. Taşeri, S. Şevik, M. Gülcü, G. Uysal Seçkin, E.C. Dolgun. 2019. Heat pump drying of grape pomace: Performance and product quality analysis. *Drying Technology* 37(14):1766–1779.

Alves-Filho, O. 2002. Combined innovative heat pump drying technologies and new cold extrusion techniques for production of instant foods. *Drying Technology* 20(8):1542–1557.

Alves-Filho, O., T.M. Eikevik. 2008. Hybrid heat pump drying technologies for porous materials. In: *Proceedings of the 16th International Drying Symposium (IDS2008)*, Hyderabad, India, November 9–12.

Alves-Filho, O., T. Lystad, T. Eikevik, I. Strømmen. 1999. A new carbon dioxide heat pump dryer: An approach for better product quality, energy use and environmentally friendly technology. Symposium on Drying Technologies 1999, St-Hyacinthe, Québec, Canada.

Alves-Filho, O., P.G. Pascual, T.M. Eikevik, I. Strommen. 2004. Dehydration of green peas under atmospheric freeze-drying conditions. In: *Proceedings of the 14th International Drying Symposium (IDS2004)*, Volume C, pp. 1521–1528, Sao Paulo, Brazil, August 22–25.

Alves-Filho, O. Tokle, T. 1999. Heat pumps dry food products in Norway. *IEA Heat Pump Centre Newsletter* 17(4):4–5.

Carrington, C.G., P. Bannister, Q. Liu. 1995. Performance analysis of a dehumidifier using HFC-134a. *International Journal of Refrigeration* 18:477–488.

Ceylan, I., M. Aktas, H. Dogan. 2007. Mathematical modelling of drying characteristics of tropical fruits. *Applied Thermal Engineering* 27:1931–1936.

Chen, G., P. Bannister, C.G. Carrington, P.T. Velde, F.C. Burger. 2002. Design and application of a dehumidifier dryer for drying pine cones and pollen catkins. *Drying Technology* 20(8):1633–1643.

Chin, S.K., C.L. Law. 2010. Product quality and drying characteristics of intermittent heat pump drying of *Ganoderma tsugae murrill*, *Drying Technology* 28:1457–1465.

Chua, K.J. 2000. Dynamic modelling, experimentation, and optimization of heat pump drying for agricultural products. Ph.D. thesis, Department of Mechanical Engineering, National University of Singapore, Singapore.

Chua, K.J., S.K. Chou, J.C. Ho, N.A. Hawlader. 2002. Heat pump drying: Recent developments and future trends. *Drying Technology* 20(8):1579–1610.

Chua, K.J., S.K. Chou, W.M. Yang. 2010. Advances in heat pump systems: A review. *Applied Energy* 87(12):3611–3624.

Colak, N., A. Hepbasli. 2005. Exergy analysis of drying of apple in a heat pump dryer. In: *Proceedings of the 2nd International Conference of Food Industries & Nutrition Division on Future Trends in Food Science and Nutrition*, pp. 145–158.

Colak, N., A. Hepbasli. 2009. A review of heat pump drying: Part 1– systems, models and studies. *Energy Conversion and Management* 50(9):2180–2186.

Eikevik, T.M., I. Strommen, O. Alves-Filho. 2005. Heat pump fluidized bed dryer with CO_2 as refrigerant: Measurements of COP and SMER. In: *Proceedings of the 3rd Nordic Drying Conference*, Karlstad, Sweden, June 15–17.

Fornasieri, F., F. Mancini, S. Minetto. 2010. Theoretical analysis and experimental test of a CO_2 household heat pump dryer. In: *Proceedings of the 9th IIR Gustav Lorentzen Conference*, Sydney, Australia.

Hawlader, M.N.A., C.O. Pereira, M. Tian. 2006. Comparison of the retention of 6-gingerol in drying under modified atmosphere heat pump drying and other drying methods. *Drying Technology* 24:51–56.

Hepbasli, A., N. Colak, E. Hancioglu, F. Icier, Z. Erbay. 2010. Exergo-economic analysis of plum drying in a heat pump. *Drying Technology* 28:1385–1395.

Honma M., T. Tamura, Y. Yakumaru, F. Nishiwaki. 2008. Experimental study on compact heat pump system for clothes drying using CO_2 as a refrigerant. In: *Proceedings of the 7th IIR Gustav Lorentzen Conference on Natural Working Fluids*, Trondheim, Norway.

Jinjiang, Z., Y. Wu. 2007. Experimental study on drying high moisture content paddy by super-conducting heat pump dryer. In: *Proceedings of the 5th Asia-Pacific Drying Conference*, Hong Kong, August 13–15.

Klöcker, K., E.L. Schmidt, F. Steimle. 2001. Carbon dioxide as a working fluid in drying heat pump. *International Journal of Refrigeration* 24(2):100–107.

Klöcker, K., E.L. Schmidt, F. Steimle. 2002. Drying heat pump using carbon dioxide as working fluid. *Drying Technology* 20(8):1659–1671.

Kuan, M., Y. Shakir, M. Mohanraj, Y. Belyayev, S. Jayaraj, A. Kaltayev. 2019. Numerical simulation of a heat pump assisted solar dryer for continental climates. *Renewable Energy* 143:214–225.

Kuzgunkaya, E.H., A. Hepbasli. 2007a. Exergetic evaluation of drying of laurel leaves in a vertical ground-source heat pump drying cabinet. *International Journal of Energy Research* 31:245–258.

Kuzgunkaya, E.H., A. Hepbasli. 2007b Exergetic performance assessment of a ground-source heat pump drying system. *International Journal of Energy Research* 31:760–777.

Lee, K.H., O.J. Kim. 2007. Experimental study on the energy efficiency and drying performance of the batch-type heat pump dryer. In: *Proceedings of the 5th Asia-Pacific Drying Conference*, Hong Kong, August 13–15.

Li, M., Y. Ma, W. Gong, W. Su. 2009. Analysis of CO_2 trans-critical cycle heat pump dryers. *Drying Technology* 27:548–554.

Liu, Y., Y. Zeng, X. Hu, X. Sun. 2020. Effect of ultrasonic power on water removal kinetics and moisture migration of kiwifruit slices during contact ultrasound intensified heat pump drying. *Food Bioprocess Technology* 13(3):430–441.

Mancini, F., S. Minetto, E. Fornasieri. 2011. Thermodynamic analysis and experimental investigation of a CO_2 household heat pump dryer. *International Journal of Refrigeration* 34:851–858.

Minea, V. 2004. Heat pumps for wood drying: New developments and preliminary results. In: *Proceedings of the 14th International Drying Symposium (IDS 2004)*, Volume B, pp. 892–899, Sao Paulo, Brazil, August 22–25.

Minea, V. 2006. Hardwood drying with low temperature heat pumps. In: *Proceedings of the 15th International Drying Symposium*, Volume C, pp. 1757–1762, Budapest, Hungary, August 20–23.

Minea, V. 2007. Energetic and ecological aspects of wood drying with heat pumps. In: *Proceedings of the 5th Asia-Pacific Drying Conference*, Volume 1, pp. 434–442, Hong Kong, 13–15 August.

Minea, V. 2008. Design and control optimizations of drying heat pumps. In: *Proceedings of the 16th International Drying Symposium (IDS 2008)*, Hyderabad, India, 9–12 November.

Minea, V. 2010a. Improvements of high-temperature drying heat pumps. *International Journal of Refrigeration* 33(1):180–195.

Minea, V. 2010b. Industrial drying heat pumps. *Refrigeration: Theory, Technology and Applications*, Nova Science Publishers, Hauppauge, New York, NY, pp. 1–70.

Minea, V. 2011. *Industrial Drying Heat Pumps in Refrigeration: Theory, Technology and Applications*, Nova Science Publishers, Hauppauge, New York, NY, pp. 1–70.

Minea, V. 2012. Efficient energy recovery with wood drying heat pumps. *Drying Technology* 30:1630–1643.

Minea, V. 2013a. Review: Drying heat pumps—Part I: System integration. *International Journal of Refrigeration* 36:643–658.

Minea, V. 2013b. Review: Drying heat pumps—Part II: Agro-food, biological and wood products. *International Journal of Refrigeration* 36:659–673.

Minea, V. 2015a. Overview of heat pump-assisted drying systems. Review—Part I: Integration, control complexity, and applicability of new innovative concepts. *Drying Technology* 33(5):515–526.

Minea, V. 2015b. Overview of heat pump-assisted drying systems. Review—Part II: Data provided vs. results reported. *Drying Technology* 33(5):527–540.

Minea, V. 2019. *Industrial Heat Pump-Assisted Wood Drying*, CRC Press, Philadelphia, PA.

Mortezapour, H., B. Ghobadian, S. Minaei, M.H. Khoshtaghaza. 2012. Saffron drying with a heat pump-assisted hybrid photovoltaic-thermal solar dryer. *Drying Technology* 28:560–566.

Mujumdar, A.S., S. Devahastin. 2000. Fundamental principles of drying. In: *Mujumdar's Practical Guide to Industrial Drying, Principles, Equipment and New Developments*, edited by S. Devahastin, Thananuch Business Ltd. Publication, Bangkok, Thailand.

Mujumdar, A.S., L.X. Huang. 2007. Global R&D needs in drying. Drying Technology 25:647–658.

Ong, S.P, Law, C.L. 2009. Intermittent heat pump drying in fruit and vegetable processing. In: *Proceedings of the 6th Asia-Pacific Drying Conference (ADC2009)*, Bangkok, Thailand, October 19–21.

Pal, U.S., Khan, Md. K. 2008a. Calculations steps for the design of different components of heat pump dryers under constant drying rate condition. *Drying Technology* 26:864–872.

Pal, U.S., M.K. Khan, S.N. Mohanty. 2008b. Heat pump drying of green sweet pepper. *Drying Technology* 226:1584–1590.

Perry, E.J. 1981. Drying by cascade heat pumps: Periodical article. Institute of Refrigeration, pp. 1–8 (https://iifiir.org/en/fridoc, accessed May 2, 2025).

Phaphvangwittayakul, W., G.S.V. Raghavan, P. Terdtoon. 2007. Loop thermosyphon application for heat pump drying. In: *Proceedings of the 3rd Inter-American Drying Conference (IADC2007)*, Montréal, Canada, August 21–23.

Queiroz, R., A.L. Gabas, R.N. Telis. 2004. Drying kinetics of tomato by using electric resistance and heat pump dryers. *Drying Technology* 22(7):1603–1620.

Rahman, M.S., C.O. Perera, C. Thebaud. 1998. Desorption isotherm and heat pump drying kinetics of peas. *Food Research International Journal* 30(7):485–491.

Saensabai, P., S. Prasertsan. 2003. Effects of component arrangement and ambient and drying conditions on the performances of heat pump dryers. *Drying Technology* 21(1):103–127.

Salehi, F. 2021. Recent applications of heat pump dryer for drying of fruit crops: A review. *International Journal of Fruit Science* 21(1):546–555.

Sarkar, J., S. Bhattacharyya, M.R. Gopal. 2006. Trans-critical CO_2 heat pump dryer: Part 2. Validation and simulation results. *Drying Technology* 24:1593–1600.

Şevik, S. 2014. Experimental investigation of a new design solar-heat pump dryer under the different climatic conditions and drying behavior of selected products. *Solar Energy* 105:190–205.

Strømmen, I., K. Kramer. 1994. New applications of heat pumps in drying processes. *Drying Technology* 12(4):889–901.

Strumillo, C. 2006. Perspectives on development in drying. *Drying Technology* 24:1059–1068.

Sunthonvit, N., G. Srzednicki, J. Craske. 2007. Effect of drying treatments on the composition of volatile compounds in dried nectarines. *Drying Technology* 25:877–881.

Velazquez, G., F. Chenlo, R. Moreira, E. Cruz. 1997. Grape drying in a pilot plant with a heat pump. *Drying Technology* 15:899–920.

Xanthopoulos, G., N. Oikonomou, G. Lambrinos. 2007. Applicability of a single layer drying model to predict the drying rate of all figs. *Journal of Food Engineering* 81:553–559.

Zielinska, M., P. Zapotoczny, P., O. Alves-Filho, T.M. Eikevik, W. Blaszczak. 2012a. Combined heat pump fluidized bed atmospheric freeze drying and microwave vacuum drying of green peas. In: *Proceedings of the 18th International Drying Symposium*, Xiamen, China, 11–15 November.

Zielinska, M., P. Zapotoczny, O. Alves-Filho, T.M. Eikevik, W. Blaszczak. 2012b. Microwave vacuum assisted drying of green peas using heat pump and fluidized bed. In *Proceedings of the 18th International Drying Symposium*, Xiamen, China, 11–15 November.

Zsivanovits, G., B. Brashlyanova, O. Karabadzhov, M. Marudova-Zsivanovits. 2013. Heat pump drying of red grape. *Journal of Food Physics* 26:19–28.

6 Optimum Coupling of Thermodynamic Systems

6.1 INTRODUCTION

Drying of wet porous and hygroscopic solid materials, which occurs by supplying heat to dried products to evaporate the liquid (water) contained inside, is a complex, highly nonlinear, and energy-intensive process. In many countries, it consumes, for example, up to 70% of all the energy used in the wood industry and 50% of the energy employed in textile and food industrial sectors (Chou and Chua 2007).

Conventional hot air convective batch drying technique is the most widely used method for production (and consumption) of dried fruits, vegetables, and woods (e.g., for building construction and furniture).

Heat pump-assisted drying systems consist of coupled slightly modified conventional hot air convective batch enclosures (also called chambers, compartments, or kilns) with air-to-air mechanical vapor compression heat pumps (acting as cooling, dehumidifying, and reheating devices). They are employed to dry various agro-food products (e.g., fruits and vegetables) at relatively low temperatures (up to 55°C) and other materials (e.g., woods) at medium and high temperatures (between 60°C and 110°C), generally in combination with fossil fuels (oil, natural gas) or renewable (biomass) fuels, or electricity as supplementary (and backup) energy sources.

Hot air convective dryers and air-to-air mechanical vapor compression heat pumps both are complex thermodynamic systems. Consequently, such coupled systems are much more complex than each of them separately. Therefore, they should not be analyzed independently due to the complex interaction between the air drying process and the heat pump (refrigerant) thermodynamic cycle. While the fundamentals of drying processes (i.e., heat and mass transfer theories) and thermodynamic cycles of heat pumps are relatively well-known, the interaction between these two thermodynamic systems requires careful theoretical and experimental approaches. Therefore, one of the most challenging tasks for heat pump-assisted drying industry is the optimum coupling of hot air convective batch dryers with air-to-air mechanical vapor compression heat pumps.

This section discusses some of technical issues related to this important challenge to provide optimum drying parameters and energy performances of heat pump-assisted drying systems.

DOI: 10.1201/9781003618386-6

6.2 ADVANTAGES AND LIMITATIONS OF HEAT PUMP-ASSISTED DRYING SYSTEMS

Despite their higher capital cost, regular maintenance for the main components, and requirements for periodical refrigerant checking and charging, the heat pump-assisted drying systems offer advantages over conventional hot air convective batch dryers (Carrington et al. 1995; Minea 2004; Minea 2006; Minea 2012) (*note:* not listed in order of importance): (i) higher energy efficiency because of the ability to recover latent and sensible heat from the hot and moist drying air; (ii) lower energy consumed for each unit of moisture (water) removed; (iii) accurate control of drying conditions; (iv) in many applications, shorter drying cycles; (v) improved final quality of certain dried materials; (vi) ability to operate independent of outside ambient air conditions; (vii) environment-friendly drying systems because no water vapor, polluted and/or toxic gases, odors, and chemical pollutants are discharged into the atmosphere; (viii) in the case of wood drying, the water condensate can be recovered and disposed of, offering chemical treatment opportunity, and, sometimes, allowing for recovery of some valuable substances; in other words, volatile organic chemicals, normally removed with the vented moist air, now appear in the condensate streams, which potentially could be sent to separate units for chemical treatment; (ix) can be adapted to conventional batch drying enclosures which are reasonably airtight, moisture-resistant, and well insulated; (x) able to better control dry- and wet-bulb temperatures, and relative humidity of drying air; (xi) waste heat recovery and heat demand occur simultaneously; (xii) may save up to 56% of operating (energy) costs compared to conventional drying methods, resulting in relatively shorter payback periods; and (xiii) large-scale industrial heat pump-assisted dryers help to use lower-pressure and lower-capacity boilers for steam generation, thereby eliminating the requirement for stationary engineers or for additional investments for new boilers, generally very expensive.

Compared to conventional hot air convective dryers, limitations of heat pump-assisted dryers include (Cech and Pfaff 1993; Minea 2010a; Minea 2010b) (*note:* not listed in order of importance): (i) higher initial capital and maintenance costs; (ii) more complex structures and operating conditions; (iii) need for qualified and competent design engineers and operating technicians for components such as heat pump compressors and blowers, heat exchangers, refrigerant, and drying air filters; (iv) additional floor spaces and/or buildings required for heat pumps' components; (v) need for temperature-resilient materials and fluids (refrigerants, oils, belts, etc.); (vi) risk of refrigerant leaks; (vii) higher consumption of electricity (a high-grade energy source) for driving the heat pump compressor(s) and blower(s); (viii) for high-temperature heat pump-assisted drying systems operating in cold climates, higher requirements for supplementary (and backup) heating sources; (ix) to prevent excessive heat losses during cold winter months, heat pump-assisted batch dryers should be well-insulated; (x) maximum dry-bulb temperatures of drying air are limited due to safe working temperatures of heat pump compressors and other thermo-sensible components; (xi) in countries where the prices of electricity are relatively high compared, for example, to those of biomass, there will be a negative effect on the overall economic profitability; and

(xii) reliability and energy performances of heat pump-assisted dryers were, during the past few decades, often disappointed due to inadequate batch dryer-heat pump coupling, operating, and control strategies.

Today, the heat pump-assisted drying technology is not worldwide spread in industrial applications as it should be, particularly in the field of drying agro-food and wood sectors. Reasons for this, among others, are (i) some potential users are not certain about the heat pumps' long-term reliability; (ii) lack of good hardware in some types of potential applications; (iii) lack of experimental and demonstration installations in different types of industries; (iv) lack of required knowledge engineering of heat pump technology in most of target industries; and (v) relatively higher costs of electricity versus those of fossil fuels affecting the commercial viability of heat pump-assisted dryers.

6.3 SUBCRITICAL NONREVERSIBLE AIR-TO-AIR MECHANICAL VAPOR COMPRESSION HEAT PUMPS

Among other methods aiming to reduce the global fossil energy consumption in drying industry, there are heat pump-assisted dryers. These systems generally integrate air-to-air nonreversible mechanical vapor compression heat pumps (acting as cooling, dehumidifying, and heating devices) with slightly modified conventional batch drying enclosures (also called chambers or compartments). Heat pumps recover sensible and latent heat by condensing moisture from a part of the moist drying air and supply it back to batch dryers to reheat the same air streams. Such a process may accelerate the drying cycles, preserve the quality of dried materials, and reduce the overall fossil energy consumption.

This section provides some basic information about air-to-air subcritical nonreversible mechanical vapor compression heat pumps that are used in most of heat pump-assisted drying systems.

6.3.1 Basic Configuration and Real Thermodynamic Cycle

Subcritical, nonreversible, air-to-air, mechanical vapor compression heat pumps are used to transfer energy from relatively low-enthalpy (temperature) reservoirs (called heat sources) to warmer thermal reservoirs (called sink sources) using physical properties of volatile working fluids (refrigerants), as well as some amount of external electrical primary energy to run the compressors, and auxiliary equipment such as drying air blowers and/or water or antifreeze liquid-circulating pumps.

Mechanical vapor compression heat pumps coupled with hot air convective batch dryers work only in heating mode according to the nonreversible subcritical Rankine reverse thermodynamic cycle represented in Figure 6.1a. It can be seen that the main components are (i) compressors (reciprocating, scroll, or screw); (ii) expansion devices (thermostatic or electronic valves, orifices, or capillary tubes); (iii) heat exchangers (condensers, superheaters, subcoolers, and evaporators); (iv) auxiliary apparatus (liquid receivers and suction accumulators); and (v) controls.

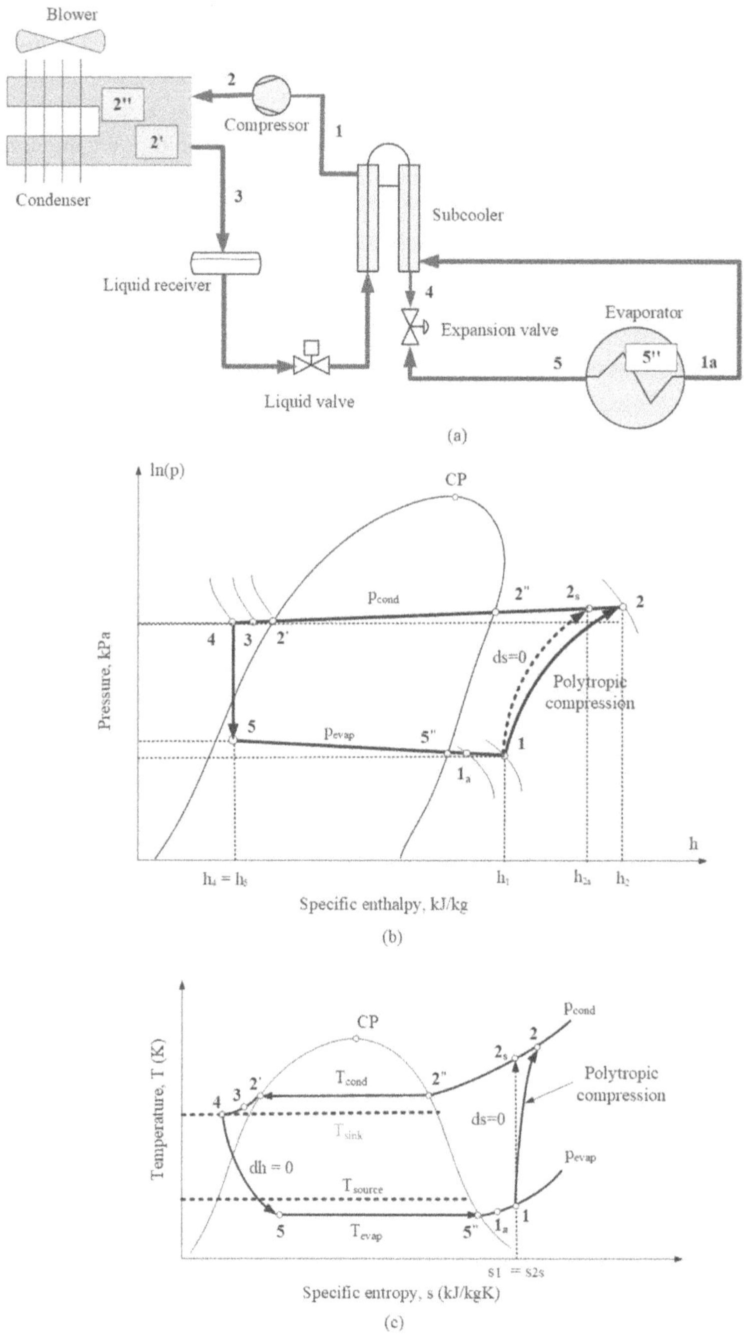

FIGURE 6.1 (a) Main mechanical components of subcritical, nonreversible (operating in heating mode only) air-to-air mechanical vapor compression heat pumps. (b) Thermodynamic cycle represented in ln(p)-h diagram. (c) Thermodynamic cycle represented in T–s diagram. cond., condensing; CP, critical point; evap., evaporating; h, specific enthalpy; p, pressure.

The main thermodynamic processes of subcritical, nonreversible air-to-air mechanical vapor compression heat pumps using pure low- or high-temperature refrigerants are represented in $\ln(p) - h$ (Figure 6.1b) and $T - s$ (Figures 6.1c) diagrams.

i. The refrigerant leaving the evaporator as superheated vapor (state 1) is submitted to polytropic (irreversible) compression process up to the superheated state 2 (process $1-2$); the electrical energy input is converted to shaft work to raise the refrigerant pressure and temperature; by increasing the vapor pressure, the condensing temperature is increased to a level higher than that of the heat source (T_{source}); if the compression process is considered ideal (adiabatic), at the end of such a process ($1-2s$), the specific enthalpy (h_{2s}) of the superheated vapor is at pressure p_{cond}, but with the same entropy as the vapor before compression ($s_{2s} = s_1$); however, the final state 2 of the real (polytropic) compression process ($1-2$) lies anywhere on the constant pressure line p_{cond}, on the right side of theoretical state point 2s.
ii. In the condenser, the superheated vapor is first desuperheated from state 2 to saturated state 2″ and, then, undergoes a two-phase condensation at constant temperature (T_{cond}) and pressure (p_{cond}) (process 2″-2′);
iii. Before leaving the condenser, the saturated liquid (state 2′) is subcooled at constant pressure to a lower temperature (process $2'-3$) in order to reduce the risks of flashing within the expansion valve; subcooling of the liquid refrigerant ensures that only liquid and no vapor bubbles enter the expansion device, allowing for better flow control and smaller devices; also, if vapor bubbles were to enter the expansion device, there would be less liquid refrigerant available in the evaporator to produce cooling, while the compressor still has to compress that portion of the vapor that did not contribute to the cooling capacity; this represents a loss of capacity and efficiency; on the other hand, too much subcooling indicates that a considerable portion of the condenser volume is filled by single-phase liquid, thus, the area available for heat rejection from the condensing fluid is smaller than it could be; as a consequence, the saturation temperature is increased along with the saturation pressure and, as a result, the compressor work increases; the degree of subcooling at the condenser outlet is primarily determined by the amount of refrigerant charged to a heat pump and the setting of the expansion device primarily determines the degree of superheat at the evaporator outlet.
iv. During desuperheating (process $2-2''$), condensing (process $2''-2'$), and subcooling (process $2'-3$), heat is rejected by the condenser to the heat sink medium (gas or liquid); within all these processes, the refrigerant exhibits pressure drops.
v. After the condenser, the liquefied subcooled refrigerant (at state 3) enters a subcooler where it is further subcooled (process 3-4) by transferring heat to the colder refrigerant coming from the evaporator at state 1a (process $1a-1$).
vi. Subcooled saturated liquid at state 4 enters the expansion valve where an expansion process at constant enthalpy (i.e., drop in pressure accompanied by a drop in temperature) takes place in order to reduce the refrigerant pressure at a level corresponding to an evaporating temperature (T_{evap}) below

to the heat source temperature (T_{source}) (process $4-5$); the expansion valve controls the refrigerant flow into the evaporator in order to ensure its complete evaporation, maintain an optimum superheat in order to avoid the liquid refrigerant to enter the compressor, and also avoid excessive superheat that may lead to overheating of the compressor.

vii. The saturated mixture (two-phase) refrigerant at state 5 enters the evaporator where it absorbs (recover) heat from the heat's source thermal carrier and undergoes change from liquid–vapor mixture to saturated vapor at constant pressure (p_{evap}) and temperature (T_{evap}) (process 5-5″); inside the evaporator, the saturated vapor (state 5″) is slightly superheated up to state $1a$ (process $5''-1a$) before entering the subcooler it where undergoes the heating process $1a-1$; the vapor superheating is desirable for two reasons. First, it assures that all the refrigerant is evaporated where it should contribute to the cooling capacity. Second, superheating usually occurs within the evaporator for the purpose of insuring that the compressor receives only vapor without any liquid entrained that could damage the compressor; however, excessive superheat may lead to overheating of the compressor.

viii. The superheated vapor at state 1 enters the compressor where a polytropic compression process 1–2 takes place; at this point, the heat pump real thermodynamic cycle restarts.

Theoretical thermal capacities of both heat pump's evaporator and condenser depends on their respective thermal capacities that are functions of refrigerant flow rates and refrigerant-side specific enthalpy changes:

$$\dot{Q}_{evap} = \dot{m}_{refr} \cdot \left(h_1 - h_5\right) \tag{6.1}$$

$$\dot{Q}_{cond} = \dot{m}_{refr} \cdot \left(h_2 - h_3\right) \tag{6.2}$$

where

$\dot{Q}_{evap}$ is the evaporator theoretical thermal power (kW);
$\dot{m}_{refr}$ is the refrigerant mass flow rate (kg/s);
h_1 is the refrigerant-specific (mass) enthalpy at the evaporator outlet (kJ/kg);
h_5 is the refrigerant-specific (mass) enthalpy at the evaporator inlet (kJ/kg);
$\dot{Q}_{cond}$ is the condenser theoretical thermal power (kW);
h_2 is the refrigerant actual specific (mass) enthalpy at the condenser inlet (kJ/kg); and
h_3 is the refrigerant-specific (mass) enthalpy at the condenser outlet (kJ/kg).

The isentropic efficiency of the actual compression process $1-2$ versus the ideal (adiabatic) process ($1-2s$, where $s_1 = s_{2s}$), is defined as follows (Figure 6.1b and c):

$$\eta_s = \left(h_{2s} - h_1\right)/\left(h_2 - h_1\right) < 1 \tag{6.3}$$

where

η_s is the isentropic efficiency of actual compression process (–);

h_{2s} is the refrigerant-specific (mass) enthalpy at the end of the ideal (adiabatic) compression process (kJ/kg);

h_1 is the refrigerant-specific (mass) enthalpy at the compressor inlet (kJ/kg); and

h_2 is the refrigerant actual specific (mass) enthalpy at the compressor outlet during the real (polytropic) compression process (kJ/kg).

Finally, the compressor theoretical (Equation 6.4) and real (Equation 6.5) electrical power input can be determined by using the following energy conservation expressions (see Figure 6.1c and d):

$$\dot{W}_{compr}^{theoretic} = \dot{m}_{refr} \cdot \left(h_{2s} - h_1\right) \tag{6.4}$$

$$\dot{W}_{compr}^{real} = \dot{m}_{refr} \cdot \left(h_2 - h_1\right) \tag{6.5}$$

where

$\dot{W}_{compr}^{theoretic}$ is the compressor theoretical electrical power input (kW);

$\dot{m}_{refr}$ is the refrigerant mass flow rate (kg/s);

h_{2s} is the refrigerant theoretical specific (mass) enthalpy at the compressor outlet (kJ/kg);

h_1 is the refrigerant-specific (mass) enthalpy at the compressor inlet (kJ/kg);

$\dot{W}_{compr}^{real}$ is the compressor real electrical power input (kW); and

h_2 is the refrigerant actual (real) specific (mass) enthalpy at the compressor outlet (kJ/kg).

6.3.2 Performance Indicators

In the heating mode, the electrically driven subcritical mechanical vapor compression heat pumps based on the ideal Carnot cycle operate between two thermal heat reservoirs having constant absolute temperatures T_{source} (heat source) and T_{sink} (heat sink), respectively (Figure 6.1c). The heating coefficient of performances for such ideal (Carnot) cycle is defined as follows:

$$COP_{Carnot}^{heating} = T_{source} / \left(T_{source} - T_{sink}\right) \approx T_{cond} / \left(T_{cond} - T_{evap}\right) \tag{6.6}$$

where

$COP_{Carnot}^{heating}$ is the Carnot (ideal) heating coefficient of performance (–);

T_{cond} is the condensing absolute temperature (K); and

T_{evap} is the evaporating absolute temperature (K).

The heating coefficient of performance of the actual (real) subcritical mechanical vapor compression heat pumps is defined as the ratio between the condenser useful (supplied) thermal power output ($\dot{Q}_{cond}$) and the heat pump total electrical power input to compressor, blower and controls, and/or liquid pumps and controls ($\dot{W}_{total}$):

$$COP_{subcritic}^{heating} = \dot{Q}_{cond} / \dot{W}_{total} \approx 1 + \dot{Q}_{evap} / \dot{W}_{total} < COP_{Carnot}^{heating} \quad (6.7)$$

where

$COP_{subcritic}^{heating}$ is the actual (real) heating coefficient of performance of subcritical mechanical vapor compression heat pump (–)

$\dot{Q}_{evap}$ and $\dot{Q}_{cond}$ are the evaporator and condenser thermal powers (kW) defined by Equations 6.1 and 6.2, respectively;

$\dot{W}_{total}$ is the heat pump total electrical power input to compressor, blower, and controls (kW); and

$COP_{Carnot}^{heating}$ is Carnot (ideal) coefficient of performance defined by Equation 6.6 (–).

In practice, the actual $COP_{subcritic}^{heating}$ of subcritical mechanical vapor compression heat pumps varies between 40% and 60% of the maximum $COP_{Carnot}^{heating}$ and depends on the difference (temperature lift) between the condensation (or heat sink) and evaporation (or heat source) temperatures. The smaller the difference, the higher the heating $COP_{subcritic}^{heating}$. For example, air-to-air (or air-to-water) subcritical mechanical vapor compression heat pumps operating in mild climates may achieve $COP_{subcritic}^{heating}$ up to 4.0, but at ambient temperatures below approximately −8°C, their $COP_{subcritic}^{heating}$ may drastically drop.

6.3.3 Low-GWP Refrigerants

Selection of refrigerants as working fluids for subcritical air-to-air mechanical vapor compression heat pumps depends on their thermodynamic properties and environmental impacts (mainly due to leakages in the atmosphere and greenhouse gas emissions associated with energy use).

To be acceptable for subcritical air-to-air mechanical vapor compression heat pump industry, the alternative refrigerants should meet the following criteria (Radermacher and Hwang 2000): (i) have properties as close as possible to those of the original HCFCs and HFCs, such as zero ozone depleting potential (ODP) and low global warming potential (GWP); (ii) be chemically inert (i.e., noncorrosive) with construction materials and fluids, such as compressor lubricants; (iii) be chemically stable in order to prevent decomposition (e.g., during high temperature and pressure at the end of compression process); (iv) be nontoxic for manufacturing and service personnel of refrigerants; (v) be nonflammable and without danger of explosion; (vi) provide high energetic efficiency; (vii) be inexpensive and easily available; (viii) have pressure–temperature properties that must fit with those of each particular application; and (ix) have relatively high enthalpy (latent heat) of vaporization because it directly impacts the compressor size and the system overall cost and energy performances.

Refrigerants as HFC-134a and HFC-410A have as main replacement the refrigerant HFO-1234yf (hydrofluoroolefin), a low-GWP substitute (GWP ≈ 4), with heat delivery temperatures limited at maximum 80°C (Table 6.1).

TABLE 6.1
Low- and Medium-Temperature Pure Refrigerants and Their Replacements

Refrigerant	ODP[a]	GWP[b]	Critical Temperature (°C)	Critical Pressure (MPa)
Present Refrigerants				
HFC-134a	0	1,430	101.08	4.0603
HFC-410A	0	2,088	72.13	4.9261
Replacement				
HFO-1234yf	0	1	94.7	3.382

[a] Based on $ODP = 1$ of CFC-11.
[b] Based on $GWP = 1$ of CFC-11 over 100 years.

To provide heat at temperatures up to 140°C with relatively high coefficients of performance, by recovering industrial waste heat at temperatures up to 80°C, environment-friendly refrigerants, such as HFC-236fa and HFC-245fa, can be used in single-stage or cascade high-temperature mechanical vapor compression heat pumps (Table 6.2).

HFC-245fa (1,1,1,3,3-pentafluoropropane) is a hydrofluorocarbon without chlorine, nearly nontoxic, with no ODP, and practically nonbiodegradable with a lifetime of 7.2 years when it eventually does escape into the atmosphere. Even if its *GWP* value is relatively high (858), HFC-245fa provides comparable and, in some cases, improve the energy performance of high-temperature heat pumps developed, for example, for industrial wood drying (Minea 2004).

As long-term options for eliminating the influence of synthetic refrigerants on global climate change, natural refrigerants, containing neither chlorine nor fluorine molecules, such as *ammonia* (NH_3, $R-717$), hydrocarbons (e.g., propane R-290), and carbon dioxide (CO_2, $R-744$) are more and more used in subcritical mechanical vapor compression heat pumps as environment-friendly alternatives. The ODPs of these natural refrigerants are zero and most of them have close to zero GWPs in comparison to HFCs (Table 6.3).

Ammonia (NH_3, R-717), an alkaline fluid with no ozone depletion potential ($ODP = 0$) and no direct global warming potential ($GWP = 0$), is suitable for usage in large industrial heat pump systems.

TABLE 6.2
High-Temperature Refrigerants for Heat Pumps

Refrigerant	ODP[a]	GWP[b]	Critical Temperature (°C)	Critical Pressure (MPa)
Water (H_2O, R-718)	0	0	374.14	22.09
HFC-236fa	0	806	124.9	3.12
HFC-245fa	0	858	154.01	3.651

[a] Based on $ODP = 1$ of CFC-11.
[b] Based on $GWP = 1$ of CFC-11 over 100 years.

TABLE 6.3
Most Promising Natural Refrigerants for Heat Pumps

Refrigerant	ODP[a]	GWP[b]	Critical Point	
			Temperature (°C)	Pressure (MPa)
Ammonia (NH_3, $R-717$)	0	0	132.25	11.483
Propane ($R-290$)	0	3	96.7	4.30
Water (H_2O, $R-718$)	0	0	373.946	22.06
Carbon dioxide (CO_2, $R-744$)	0	1	30.98	7.474

[a] Based on $ODP=1$ of CFC-11.
[b] Based on $GWP=1$ of CFC-11 over 100 years.

Hydrocarbons, such as propane ($R-290$), can be alternative refrigerants to halogenated substances since they (i) are available in nature at relatively low costs; (ii) have excellent thermodynamic and transport properties; (iii) have zero ozone depletion potential ($ODP=0$) and insignificant direct global warming potential ($GWP=3$); (iv) are miscible with mineral oils; and (v) are compatible with most of common materials. The following are the limitations of hydrocarbons: (i) high flammability with relatively low limits; (ii) precautions are required to prevent ignition in case of massive leakage; and (iii) regular monitoring and maintenance measures (as gas sensors and air removal devices) are required.

Because of its negligible impact on climate change (no ozone depletion potential and negligible direct GWP when used as a refrigerant in closed cycles), carbon dioxide (CO_2, R-744) is a promising natural refrigerant for usage in sub- and transcritical heat pumps. It is a nontoxic fluid at low concentrations (but can be harmful in higher concentrations), nonflammable, colorless, odorless, noncorrosive, heavier than air, relatively inexpensive, compatible with most of common lubricants, with excellent thermo-physical and transport properties. As a refrigerant, CO_2 operates with significantly higher pressures (up to 15 MPa on the high-pressure side) than other refrigerants, which require stronger materials and/or larger wall thicknesses.

6.4 HOT AIR CONVECTIVE BATCH DRYERS

Laboratory- and industrial-scale driers are classified according to the following criteria: (i) type of drying unit (e.g., batch or continuous, tray, rotating drum, fluidized bed, spray, etc.); (ii) physical form of the dried material (e.g., fibrous, powder, granules, slurry, etc.); and (iii) methods of heat transfer (e.g., basically, convection, conduction, radiation, or dielectric).

About 85% of laboratory- and industrial-scale dryers are forced hot air convective-type batch dryers using fossil fuels (as oil, natural gas, and biomass) and/or electricity as primary energy sources. In developed countries, they may consume up to 25% of the national industrial energy demands. Most heat losses are due to the moist exhaust air and the poor thermal insulation of drying enclosures. Because of heat losses, historically, attention has been focused on reducing and/or recovering waste heat as part of the global effort to reduce energy consumption and control greenhouse gas emissions.

Optimum selection of energy-efficient dryers represents a compromise between the initial capital and operating costs and depends on (i) properties of the products to dry; (ii) relative cost of fossil fuels and electricity; and (iii) ability to efficiently control the drying processes; and production throughput.

Compared to natural (yard) air drying, hot air convective batch dryers presents the following advantage (Cech and Pfaff 1993): (i) are more efficient because they reduce the drying times by using higher drying temperatures; and (ii) better control of final moisture contents of dried materials regardless of weather conditions.

The following are the disadvantages of conventional batch dryers: (i) much higher capital, operation, and maintenance costs as compared to natural (yard) air drying; and (ii) heat losses and inefficiencies in heat transfer may limit the drying dry-bulb temperatures to between 82°C and 93°C.

6.4.1 Typical Construction

Industrial-scale conventional hot air convective batch driers (as those schematically shown in Figure 6.2a and b) are large building (also called chambers, enclosures, and rooms) that provide artificial environments allowing for optimizing the products' drying processes.

The dryers' geometry and construction should reduce energy requirements (mainly, for heating and fans), shorten the drying times, and provide more uniform moisture contents throughout the stacks of dried materials.

Industrial-scale driers are usually available in variety of sizes and shapes (e.g. capacities between 10 m^3 and 200 m^3), constructed of sheet steel metal airtight and thermally insulated panels in order to reduce the energy consumption, prevent condensation on the walls, and for better control of air leaks out of the structure. They are designed to supply heat and control the air dry- and wet-bulb temperatures, relative and absolute humidity, and circulation (mass flow rate, velocity). The control modes can be manual, semi-, or fully automatic, depending on the dryer size and material to dry. Generally, the conventional batch dryers are charged with products having similar drying characteristics, as density, permeability, anatomical properties, thickness, and initial moisture contents.

Low-temperature and medium-temperature dryers operate at standard temperatures between 20–45°C and 45–82°C, respectively, where interior fans produce air velocities of 2.5–3.6 m/s. High temperatures (83–149°C) are usually used in convective batch wood dryers to provide faster drying processes with air velocities between the boards higher than 5 m/s. Capital investments per unit capacity of high-temperature hot air convective batch dryers are higher, but drying times and energy consumption may be reduced by up to 50%.

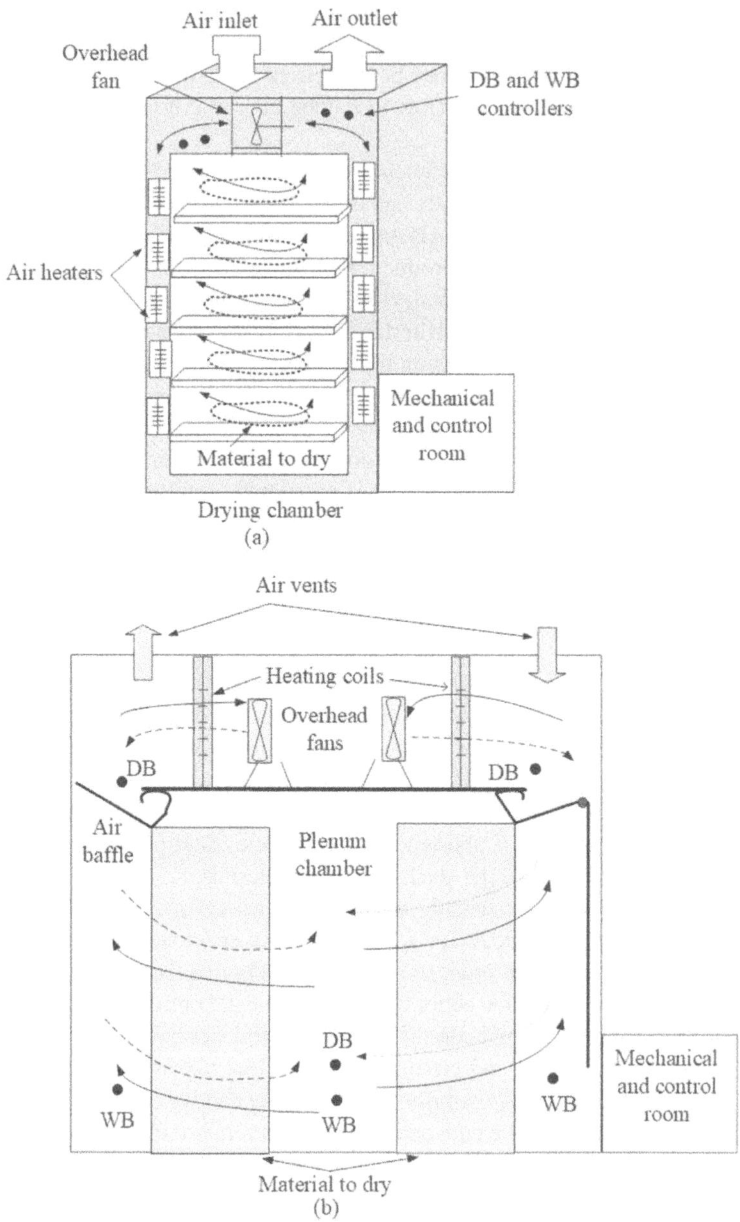

FIGURE 6.2 Schematic front view of conventional hot air convective batch dryers. (a) Typically used for drying fruits and vegetable. (b) Typically used for drying lumber boards. *Note*: Schematics not to scale; not all components shown). DB, dry-bulb temperature sensor; WB, wet-bulb temperature sensor.

6.4.2 Heating Means

Heat for drier warming-up and material supplementary (and backup) heating can be supplied to batch dryers either directly (by using combustion gases or, occasionally, electrical heating elements) or indirectly, generally by using steam, hot water, or oil flowing through plain or fined heat exchangers at temperatures of drying medium ranging from 60°C to 150°C, depending on the material to be dried.

Typically, copper (the best conductor, but expensive and subject to damage), iron (the most rigid and serviceable but subject to corrosion), or aluminum (an excellent heat conductor, but much more subject to damage) are used in finned heat exchangers.

6.4.3 Circulation of Drying Air

Forced air circulation is required to carry heat from the heating coils to the material being dried, and to carry away the moisture vaporized from the dried product. For that, the batch dryers are equipped with overhead fans powered by electrical motors which may be installed outside the drying chamber, or inside it. They have to overcome the pressure drops accompanying the changes in air direction, the losses at the inlet and outlet faces of the stack(s), and the pressure loss in forcing the air through the passageways in the material stack(s). Dryers' fans are usually situated above the material stack(s) in false ceiling in such a way as to ensure even distribution of drying air and heat needed for drying (Figures 6.2a and b).

Airflow velocity over the surface of dried materials should be sufficiently high in order to match the rate at which the moisture moves from the interior to the surface of the dried material with the vaporization rate at their surfaces, avoid the formation of death zones and, thus, provide uniform drying. Operating (energy) costs of dryer's overhead fans can be significantly reduced by using speed controllers (frequency inverters or multiple-speed fan motors). Usually, the circulation of drying air is periodically reversed to provide more effective and more uniform heating to dried materials. Airflow reversals at every few hours are used to ensure that the variation in moisture content in the airflow direction is controlled to minimize the dryer-wide variations in moisture content from material piece to material piece, and trim the excessive over- and underdrying of the outer portions of the material stack. Reversing the direction of the airflow has the effect of enhancing the drying of the wetter zone of the material stack(s).

6.4.4 Air Venting

As the hot air passes through the material stack(s), the wet material absorbs heat. It results in the movement of moisture to the materials' surfaces, where it is absorbed by the circulating drying air. When the circulating air becomes saturated with moisture, the air vents (usually, included in *roof panels*) open, and the excess moisture is vented to the atmosphere. Fresh air is then drawn into the dryer, heated, humidified, and the material drying process continues.

The air vents are arranged in one or two-vent lines, depending on the fan arrangement, along the length of dryers. Each line of air vents is automatically opened and closed by electrically or pneumatically powered motors activated by the recording–controlling instruments. Generally, cool dry air is introduced at one end of the dryer while warm moist air is expelled at the other end by using the pressure difference generated by the reversible fans. Usually, the air vent system in any dryer exhausts more air than it draws into the dryer, to accommodate the expansion in volume of incoming air as it is heated.

6.4.5 Humidification

The requirements for final quality of dried materials limit the rise to air dry-bulb temperature inside the dryers. Excessively hot air can almost completely dehydrate the materials' surfaces, so the material pores shrink. In order to avoid excessive surface dehydration of sliced fruits and vegetables, or lumber board deformations, a humidification process is usually required to compensate for humidity losses due to uncontrolled air leakages and, thus, to keep the relative humidity inside the dryer from dropping too low during the drying cycle.

The control of humidity may be accomplished by means of opening air vents or through direct steam injection via spray pipes or cold water misting.

6.4.6 Basic Controls of Hot Air Convective Batch Dryers

In hot air convective batch dryers, the heating systems ensure that the desired temperatures exist in drying chambers. It must offset energy losses due to evaporation, increases in material temperature, the heating up of the fresh air intake required to adjust the moisture content in the drying chamber, and energy lost through drying enclosure walls and doors. Its ability to offset energy loss depends on its power and on drying air velocity. The indirect heating systems (with steam, hot oil, or hot water) via heat exchangers generally provide the most effective distribution of heat and best control.

The control of hot air convective batch dryers mainly focus on changes in material average moisture content (that must be periodically measured and compared with set point values) in order to obtain the final value as close as possible to required target that depends on (i) dryer's capacity; (ii) initial moisture content of material to dry; and (iii) variation of drying air parameters (e.g., dry- and wet-bulb temperatures, and relative and absolute humidity) during the entire drying process.

However, the control based only on moisture content measurement are not reliable in industrial practice mainly because the moisture content varies from one dried piece to another. It can be measured only for a limited range of values. Therefore, it is difficult to compare the measured parameters with set point values. Ideally, once each target is reached, the drying process advances to the next step, and new set points are established for dry-bulb and wet-bulb temperatures.

The basic control system shown in Figure 6.3 contains a *PID* (proportional integral and derivative) regulator that controls auxiliary parameters such as the temperature of drying air, moisture content and mass of material stack(s), or drying time. Temperature drop across the load (*TDAL*) measures dry- and wet-bulb temperature variations and then, induces changes in dryer's temperature settings.

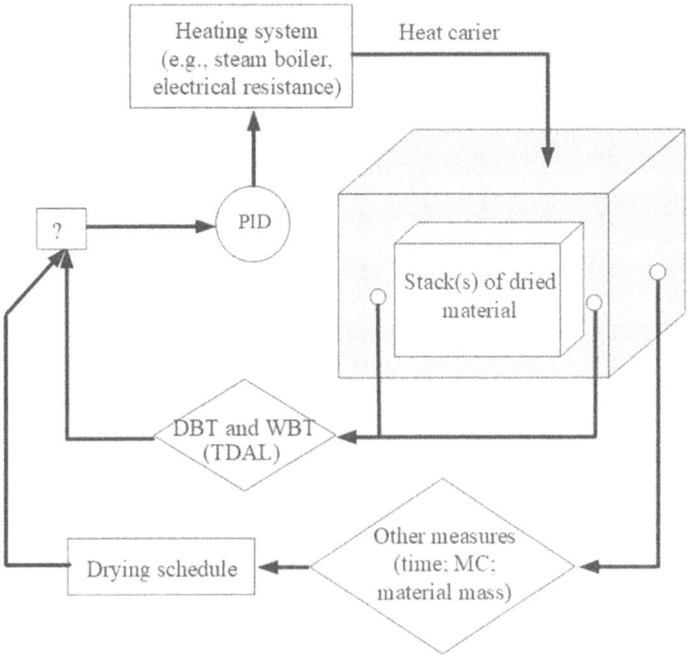

FIGURE 6.3 Simplified diagram of a typical PID drying process control. *DBT*, dry-bulb temperature; *MC*, moisture content; *PID*, proportional–integral-derivative. *TDAL*, temperature drop across the load; *WBT*, dry-bulb temperature. (Redrawn from Léger and Amazouz 2003.)

Depending on the detected errors, the PID controller regulates the heating system by raising or lowering the heat carrier fluid output (e.g., steam) from the heating source. Such a control rule can achieve results that are accurate, fast, stable, and reliable drying cycles. The *PID* control loop compares the actual air dry- and wet-bulb temperatures, or *TDAL* measurements with their respective set points, and, according to the proportional action, opens and/or shut-off a tap proportionally to the observed discrepancy between the actual parameters in drying chamber and desired values. When a set value (e.g., dry- or wet-bulb temperatures, drying air relative or absolute humidity, dew point, or elapsed drying time) is reached, the *PID* controller changes the temperatures in the drying enclosure via the heat carrier (Léger and Amazouz 2003).

6.5 OPTIMUM COUPLING OF CONVENTIONAL HOT AIR CONVECTIVE BATCH DRYERS WITH AIR-TO-AIR MECHANICAL VAPOR COMPRESSION HEAT PUMPS

Hot air convective batch dryers coupled with air-to-air mechanical vapor compression heat pumps become more complex systems due to the close interdependency of their components and thermodynamic parameters (Minea 2019; Mujumdar and Huang 2007).

Heat pump-assisted batch dryers are totally closed systems which generally use single-stage air-to-air heat pumps to condense moisture vaporized from various porous solid materials and re-warm a part of the recirculated drying air (Chua et al. 2002; Strumillo 2006).

Such advanced drying systems use supplementary (and backup) energy sources for material warming-up (preheat), such as fossil fuels (e.g., oil, propane, and natural gas) or renewables (e.g., biomass, bark) (supplied via conventional boilers) and electricity (supplied via resistive elements).

Optimum coupling of the two thermodynamic components (hot air convective batch dryers and air-to-air mechanical vapor compression heat pumps) is a challenging task for academia and industry heat pump-assisted drying communities. Their efficient operation and control rules, as well as energy performances are highly dependent on how the key components are physically coupled and thermodynamically integrated to provide normal thermodynamic conditions especially to reciprocating compressors of air-to-air mechanical vapor compression heat pumps.

Figure 6.4 shows the main steps of a standard analysis aiming to decide to use or not heat pump-assisted drying method.

6.5.1 Coupling Batch Dryers with Single Compact Mechanical Vapor Compression Heat Pumps

In hot air convective batch drying chambers coupled with single compact air-to-air mechanical vapor compression heat pumps, the heat pump evaporator and condenser are installed in the same cabinet (mechanical room) located outside (adjacent) the drying chamber (Minea 2006; Minea 2011) (Figure 6.5).

The optimum coupling arrangement shown in Figure 6.5a makes it possible to efficiently control the heat pump evaporator (thermal power) dehumidification rate, as well as the temperature and relative humidity of hot and dry air supplied to the drying chamber at state D using a constant speed blower (B).

As can be seen in Figure 6.5a and b, the warm and humid air coming from the drying chamber is divided into two flows. The first one (at state i) passes through the evaporator finned heat transfer surface where it is sensibly cooled (process $i-a$) and then dehumidified across the evaporator (process $a-b$), while the second one (at state i) bypasses the evaporator external heat transfer surface and then is mixed with the dehumidified air (state b) resulting in drying air at state mix. The mixed drying air (at state mix) passes through the external finned surface of heat pump condenser where it is heated up at constant absolute humidity (process $mix-d$). Prior to returning to the drying enclosure (that plays the role of a drying air plenum), an additional heat exchanger *HEX* allows controlling the final temperature and relative humidity of drying air that returns the drying chamber at state D (process $d-D$). Inside drying chamber, moisture (water vapor) is added to the drying air at state d, and thus its thermodynamic state changes *adiabatically* along a constant enthalpy line *(process* $\mathrm{d-i}$*)*. The dry-bulb temperature of the drying air decreases from $DBT-d$ to $DBT-i$, and its absolute humidity increases from ω_d to ω_i.The drying air is cooled up to its dew point (DEW) (process $i-a$), and then it is dehumidified (process $a-b$) and some of the water vapor will condense and leave the air stream.

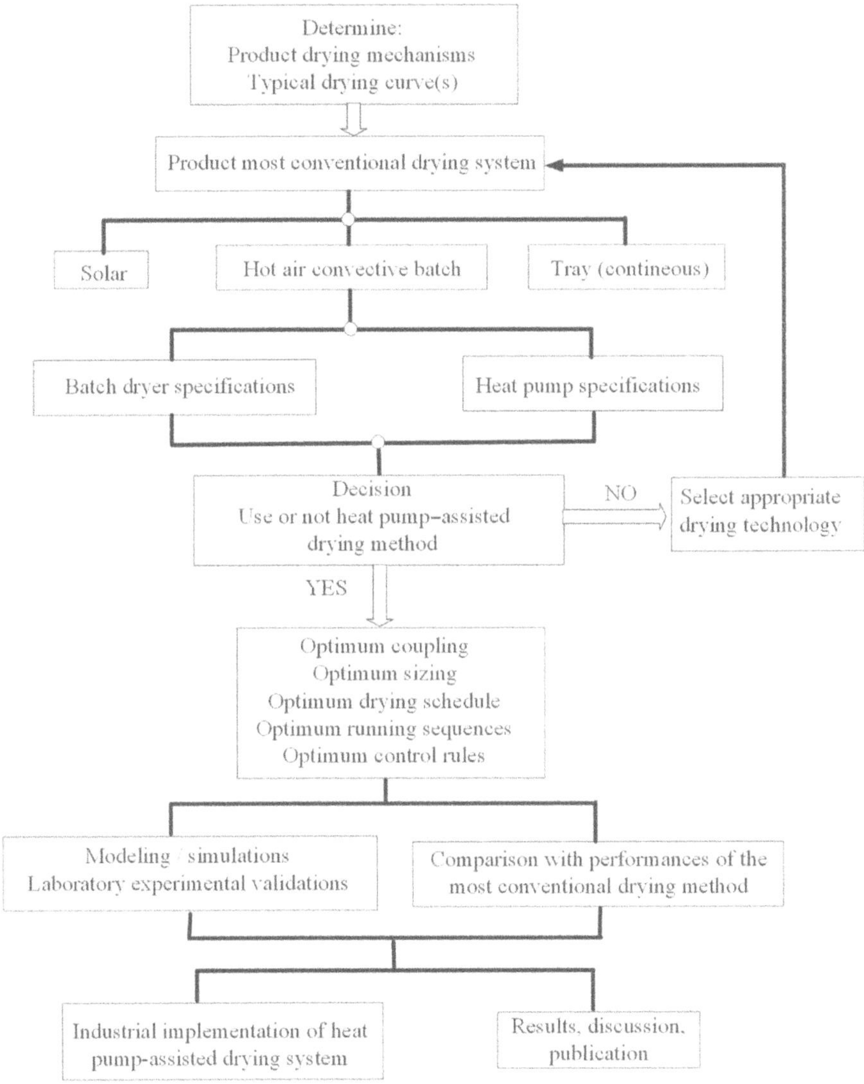

FIGURE 6.4 Main steps for selection and analysis of heat pump-assisted drying.

Cooling and dehumidifying processes involve both sensible and latent heat transfer, where sensible heat transfer is associated with the decrease in dry-bulb temperature from $DBT - i$ to DEW, and the latent heat transfer is associated with the decrease in absolute humidity from $\omega_i = \omega_a$ to ω_b. Both processes result in reductions of both dry-bulb temperature and the absolute humidity.

On the drying air side, the bypass circuit around the heat pump evaporator is recommended to vary the drying air flow through the evaporator in order to (i) control the dehumidification rate during the whole drying process; and (ii) mix the drying air leaving the evaporator at 80–90% relative humidity with the drying air leaving

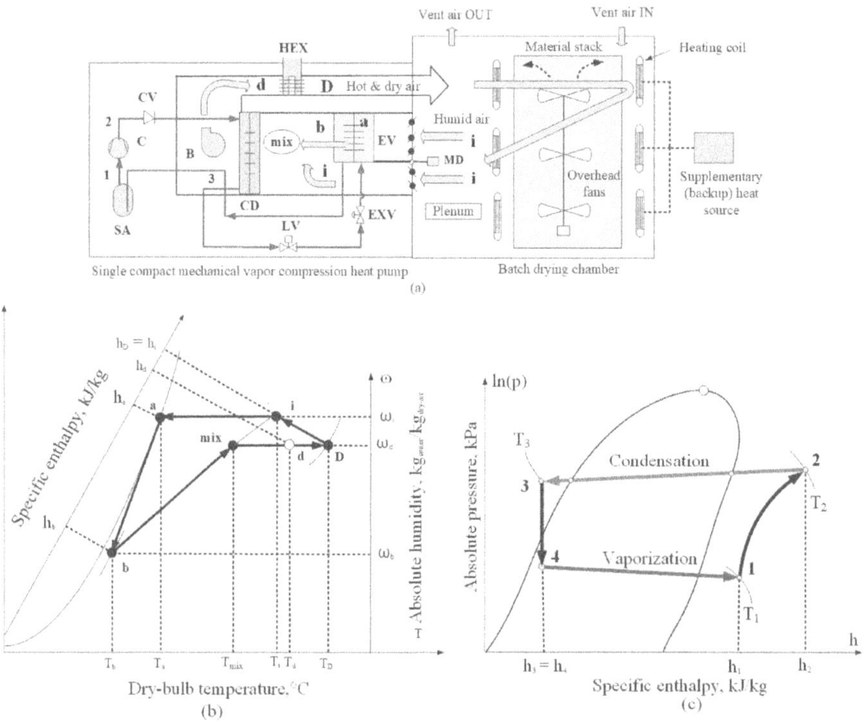

FIGURE 6.5 Batch drying chamber coupled with a single compact mechanical vapor compression heat pump. (a) Schematic. *Note:* Schematic not to scale; not all components shown. (b) Thermodynamic processes of drying air in the Mollier $(h-T)$ diagram. (c) Thermodynamic processes of heat pump refrigerant represented in ln(p)-h diagram. *B*, constant speed drying air blower; *C*, compressor; *CD*, heat pump remote condenser; *EV*, evaporator; *EXV*, expansion valve; *HEX*, heat exchanger for control of drying air final supply temperature and relative humidity; *LV*, refrigerant liquid valve; *MD*, motorized damper.

the dryer; this bypass supplies a sufficient drying air flow rate through the condenser to avoid excessive high compressor discharge pressures and temperatures; at the same time, the mixing process $(b-mix-i)$ allows the condenser to reach the desired air supply temperature without oversizing its heat transfer surface.

The variable air bypass around the heat pump evaporator is performed by the aid of motorized dampers (*MD*) that help to vary the drying air flow rate (and velocity), and, thus, the heat pump dehumidification capacity during the entire drying cycle. This is necessary since the heat pump dehumidification capacity (i.e., the evaporator thermal power) is always higher at the beginning (when the moisture content of drying air is high) and lower at the end of drying processes (when the moisture content of drying air is much lower).

The correlation between the evaporator air flow rate and the condenser air flow rate can be expressed as follows:

$$\dot{m}_{air,evap} = \alpha \cdot \dot{m}_{air,cond} \quad (6.8)$$

where

$\dot{m}_{air,evap}$ is the evaporator air flow rate (kg/s);
α is the bypass factor that varies between 0 and 1; and
$\dot{m}_{air,cond}$ is the condenser air flow rate (kg/s).

If the drying air mass flow rate is $\dot{m}_{drying\,air}$, the mass flow rate of moisture (water) added to moist air can be calculated as follows:

$$\dot{m}_{water} = \dot{m}_{drying\,air} \cdot (\omega_i - \omega_d) \tag{6.9}$$

where

$\dot{m}_{water}$ is the mass flow rate of moisture (water) added to drying air (kg_{water}/s);
$\dot{m}_{drying\,air}$ is the mass flow rate of drying air ($kg_{drying\,air}/s$);
ω_i is the absolute humidity of drying air at the outlet of drying chamber ($kg_{water}/kg_{dry\,air}$); and
ω_d is the absolute humidity of drying air at the inlet of drying chamber ($kg_{water}/kg_{dry\,air}$).

Although the actual process path will vary considerably depending on the type of surface, surface temperature, and flow conditions, the heat and mass transfer can be expressed in terms of the initial (i) and final (b) states. Since both sensible and latent heats are removed from the moist drying air, to calculate the energy required to accomplish the change, it is most convenient to use the enthalpy difference $(h_i - h_b)$ between the entering and leaving air conditions.

The moisture mass rate and thermal power balance equations for the entire process $i - a - b$ are as follows (McQuiston and Parker 1988):

$$\dot{m}_{dry\,air} \cdot \omega_i = \dot{m}_{dry\,air} \cdot \omega_b + \dot{m}_{water} \tag{6.10}$$

$$\dot{m}_{dry\,air} \cdot h_i = \dot{m}_{dry\,air} \cdot h_b + \dot{Q}_{i-b} + \dot{m}_{water} \cdot h_{fg} \tag{6.11}$$

Thus,

$$\dot{m}_{water} = \dot{m}_{dry\,air} \cdot (\omega_i - \omega_b) \tag{6.12}$$

$$\dot{Q}_{i-b} = \dot{m}_{dry\,air} \cdot \left[(h_i - h_b) - (\omega_i - \omega_b) \cdot h_{fg} \right] \tag{6.13}$$

where

$\dot{m}_{dry\,air}$ is the mass flow rate of drying air ($kg_{dry\,air}/s$);
ω_i is the absolute humidity of drying air leaving the drying chamber (entering heat pump evaporator) ($kg_{water}/kg_{dry\,air}$);
ω_b is the absolute humidity of drying air leaving heat pump evaporator ($kg_{water}/kg_{dry\,air}$);
$\dot{m}_{water}$ is the moisture flow rate extracted from the moist drying air (kg_{water}/s);
$\dot{Q}_{i-b}$ is the total heat transfer rate (or thermal power) of heat pump evaporator (kW);

h_i is the specific enthalpy of drying air at heat pump evaporator inlet ($kJ/kg_{dry\,air}$); and

h_{fg} is the specific water evaporation enthalpy (latent heat) (kJ/kg_{water}).

By referring again to Figure 6.5b, the sensible thermal power recovered can be expressed as follows:

$$\dot{Q}_{i-a} = \dot{m}_{dry\,air} \cdot \left(h_i - h_a\right) = \dot{m}_{dry\ air} \cdot \overline{c}_{p,dry\,air} \cdot (DBTi - DEW) \tag{6.14}$$

where

$\dot{Q}_{i-a}$ is the sensible thermal power recovered (kW);

$\dot{m}_{dry\,air}$ is the mass flow rate of drying air (kg_{water}/s);

h_i is the specific enthalpy of drying air at heat pump evaporator inlet ($kJ/kg_{dry\,air}$);

h_a is the specific enthalpy of drying air at its dew point ($kJ/kg_{dry\,air}$);

$\overline{c}_{p,dry\ air}$ is the drying air specific heat at constant pressure (kJ/kgK);

DBTi is the drying air dry-bulb temperature at the evaporator inlet (°C); and

DEW is the drying air dew temperature (°C).

Similarly, by neglecting the condensate energy, the latent heat transfer is given by the following:

$$\dot{Q}_{a-b} = \dot{m}_{dry\ air} \cdot \left(h_a - h_b\right) = \dot{m}_{dry\ air} \cdot \left(\omega_a - \omega_b\right) \cdot h_{fg} \tag{6.15}$$

where

$\dot{Q}_{a-b}$ is the latent thermal power (evaporator dehumidification capacity) extracted from the drying moist air (kW);

$\dot{m}_{dry\ air}$ is the mass flow rate of drying air (kg_{water}/s);

h_a is the specific enthalpy of drying air at its dew point ($kJ/kg_{dry\,air}$);

h_b is the specific enthalpy of drying air at heat pump evaporator outlet ($kJ/kg_{dry\,air}$);

ω_a is the absolute humidity of drying air at its dew point ($kg_{water}/kg_{dry\,air}$);

ω_b is the absolute humidity of drying air at heat pump evaporator outlet ($kg_{water}/kg_{dry\,air}$); and

h_{fg} is the water-specific enthalpy (latent heat) of vaporization ($kJ/kg_{dry\,air}$).

The total heat transfer rate extracted from the moist air will be

$$\dot{Q}_{i-b} = \dot{Q}_{i-a} + \dot{Q}_{a-b} = \dot{Q}_{sens} + \dot{Q}_{latent} \tag{6.16}$$

where

$\dot{Q}_{i-b}$ is the total heat transfer rate (or thermal power) extracted from the drying hot and moist air (kW);

$\dot{Q}_{i-a} = \dot{Q}_{sens}$ is the sensible thermal power extracted from the drying hot and moist air (kW); and

$\dot{Q}_{a-b} = \dot{Q}_{latent}$ is the latent thermal power extracted from the drying moist air (kW).

The adiabatic mixing process of drying air at states b and i (Figure 6.5b) is governed by the following balance equations (McQuiston and Parker 1988):

Mass balance:

$$\dot{m}_{dry\,air,i} + \dot{m}_{dry\,air,b} = \dot{m}_{dry\,air,mix} \tag{6.17}$$

Energy balance:

$$\dot{m}_{dry\,air,i} \cdot h_i + \dot{m}_{dry\,air,b} \cdot h_b = \dot{m}_{dry\,air,mix} \cdot h_{mix} \tag{6.18}$$

Moisture (water vapor) balance:

$$\dot{m}_{dry\,air,i} \cdot \omega_i + \dot{m}_{dry\,air,b} \cdot \omega_b = \dot{m}_{dry\,air,mix} \cdot \omega_{mix} \tag{6.19}$$

From Equation 6.18,

$$\dot{m}_{dry\,air,i} / \dot{m}_{dry\,air,b} = \left(h_{mix} - h_b\right) / \left(h_i - h_{mix}\right) = \left(\overline{mix - b}\right) / \left(\overline{i - mix}\right) \tag{6.20}$$

From Equation 6.19,

$$\dot{m}_{dry\,air,i} / \dot{m}_{dry\,air,b} = \left(\omega_{mix} - \omega_b\right) / \left(\omega_i - \omega_{mix}\right) = \left(\overline{mix - b}\right) / \left(\overline{i - mix}\right) \tag{6.21}$$

According to Equations 6.20 and 6.21, the state point of the drying air mixture (mix) lies on the straight line connecting the state points of the two streams being mixed (i and b) and divides the line into two segments inversely proportional to the ratio of mass flow rates of the two streams.

The drying air is finally heated at constant absolute humidity ($\omega_{mix} = \omega_d$) from state mix to state d by flowing over the heat pump refrigerant condenser (Figure 6.5a and b).

Under steady-state flow conditions, the energy balance of this heating process is as follows:

$$\dot{m}_{dry\,air} \cdot h_{mix} + \dot{Q}_{cond} = \dot{m}_{dry\,air} \cdot h_d \tag{6.22}$$

where

$\dot{m}_{dry\,air}$ is the mass flow rate of drying air (kg_{water}/s);

h_{mix} is the specific enthalpy of mixed drying air ($kJ/kg_{dry\,air}$);

$\dot{Q}_{cond} = \dot{Q}_{mix-d}$ is the sensible thermal power supplied by heat pump condenser to drying air (kW); and

h_d is the specific enthalpy of mixed drying air leaving the heat pump condenser ($kJ/kg_{dry\,air}$).

For steady-state flow conditions, the required thermal power addition is a positive number expressed as follows (McQuiston and Parker 1988):

$$\dot{Q}_{cond} = \dot{Q}_{mix-d} = \dot{m}_{dry\,air} \cdot \left(h_d - h_{mix}\right) = \dot{m}_{dry\,air} \cdot \overline{c}_{p,drying\,air} \left(T_d - T_{mix}\right) \tag{6.23}$$

where

$\dot{Q}_{mix-d}$ is the thermal power supplied by heat pump condenser to drying air (kW);
$\dot{m}_{dry\,air}$ is the mass flow rate of drying air (kg_{water}/s);
h_d is the specific enthalpy of mixed drying air leaving the heat pump condenser (entering the drying chamber) ($kJ/kg_{dry\,air}$);
h_{mix} is the specific enthalpy of mixed drying air ($kJ/kg_{dry\,air}$);
$\overline{c}_{p,dry\,air}$ is the specific heat at constant pressure of drying air ($kJ/kg \cdot K$);
T_d is the dry-bulb temperature of drying air entering the drying chamber (°C);
T_{mix} is the dry-bulb temperature of drying air at heat pump condenser inlet (°C).

The thermal power required by heat exchanger *HEX* can be expressed as follows:

$$\dot{Q}_{HEX} = \dot{m}_{dry\,air} \cdot \left(h_D - h_d\right) \tag{6.24}$$

where

$\dot{Q}_{HEX}$ is the thermal power of heat exchanger *HEX* (kW);
$\dot{m}_{dry\,air}$ is the mass flow rate of drying air (kg_{water}/s);
h_D is the specific enthalpy of drying air leaving the heat exchanger *HEX* (entering drying chamber) ($kJ/kg_{dry\,air}$); and
h_d is the specific enthalpy of mixed drying air leaving heat pump condenser (entering heat exchanger *HEX*) ($kJ/kg_{dry\,air}$).

6.5.2 Coupling Parallel Batch Dryers with Single Compact Mechanical Vapor Compression Heat Pumps

Some industries use parallel drying chambers (with their own control devices and operating strategies) for drying different types of products.

Such parallel drying chambers operating simultaneously at different drying air temperature and humidity conditions can be coupled with single *compact* mechanical vapor compression heat pumps where two air streams coming from two separate drying chambers pass over two separate evaporators of the same heat pump (Figure 6.6). Each drying chamber can, thus, operate independently and simultaneously with different types of materials, reducing energy consumption and improving industrial productivity. In such a concept, the subcooled refrigerant liquid leaving the heat pump condenser at state 3 is split into two streams. One stream enters the expansion valve $EXV-1$ of the low-pressure (temperature) evaporator $EV-LP$, and the second enters the expansion valve $EXV-2$ of the high-pressure (temperature) evaporator $EV-HP$. A high-pressure regulating valve (*PRV*) keeps the pressure of the refrigerant in the high-pressure evaporator $EV-HP$ at a higher pressure compared to that of low-pressure evaporator $EV-LP$. A check valve (*CV*), located on the suction line coming from the low-pressure evaporator $EV-LP$, prevents the high-pressure vapor coming from the high-pressure evaporator $EV-HP$ from entering the low-pressure evaporator $EV-LP$ during compressor running periods and off cycles. The mixed refrigerant at state 1 returns to the compressor where it is compressed and becomes high-pressure, high-temperature superheated vapor (process $1-2$).

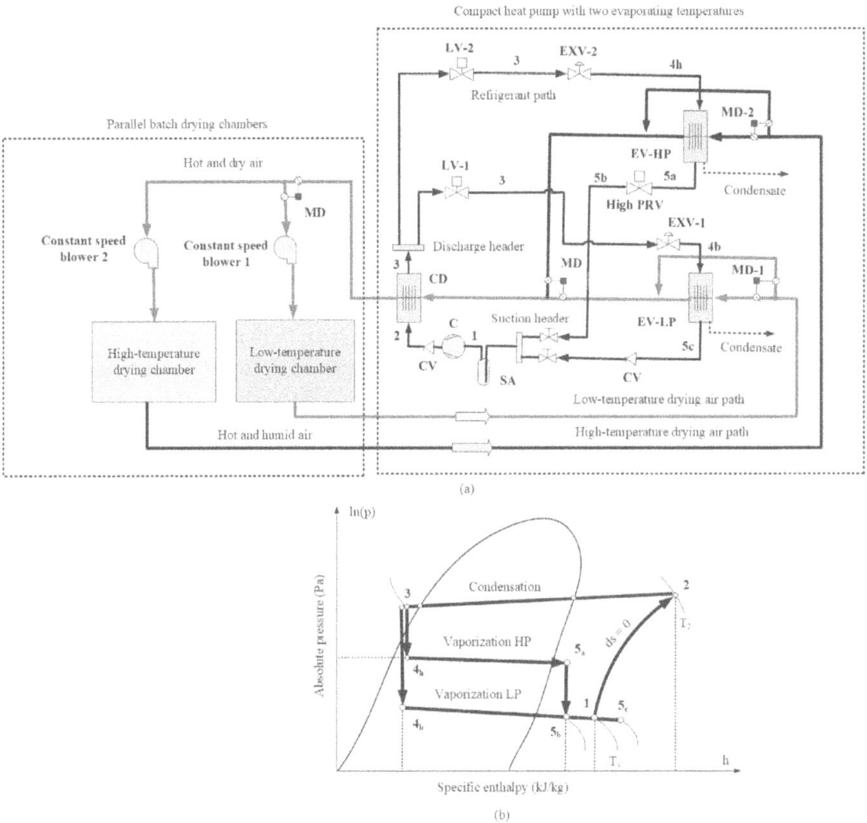

FIGURE 6.6 Parallel hot air convective batch dryers coupled with a *single compact* mechanical vapor compression heat pump. (a) Schematic. *Note:* Schematic not to scale; not all components shown. (b) Thermodynamic processes of heat pump refrigerant represented in ln(p)-h diagram. C, compressor; CD, condenser; CV, check valve; EV, evaporator; EXV, expansion valve; F, fan; $EV-HP$, high-pressure evaporator; $EV-LP$, low-pressure evaporator; LV, liquid valve; MD, motorized damper; PRV, pressure regulating valve; SA, suction accumulator.

By appropriately adjusting the expansion valves $EXV-1$ and $EXV-2$, the saturated evaporating pressures (and temperatures) are fitted to simultaneously provide drying at low temperature (in low-temperature drying chamber) and high temperature (inside the high-temperature drying chamber) for two different materials (Minea 2011).

6.5.3 Coupling Batch Dryers with Single-Split Mechanical Vapor Compression Heat Pumps

In hot air convective batch dryers coupled with *single-split* mechanical vapor compression heat pumps (Figure 6.7a), the compressors, evaporators, variable speed blowers, and controls are installed in separate (adjacent) mechanical room(s), while the remote condenser(s) is (are) installed inside the drying enclosures. Overhead fans are located above false ceilings. Their rotation senses change every 1, 2, *or* 3 hours to achieve as much as possible uniform drying conditions through the material stack(s). Air vents open when the overhead fans change their rotation direction and also when the actual dry-bulb temperature exceeds the set point. Generally, electrical or oil-burned boilers supply saturated steam to dryers' heating coils to preheat the material stacks and, when required, to provide supplemental (and backup) heat (Lewis 2003; Minea 2010a; Minea 2010b; Minea 2011).

Hot and humid air from the drying enclosure enters the heat pump evaporator at state i where it is first cooled (process $i-a$) and, then, dehumidified (process $a-b$) (Figure 6.7b). The sensible and latent heat recovered is transferred to a liquid–vapor refrigerant mixture that boils at constant, low pressure, and low temperature

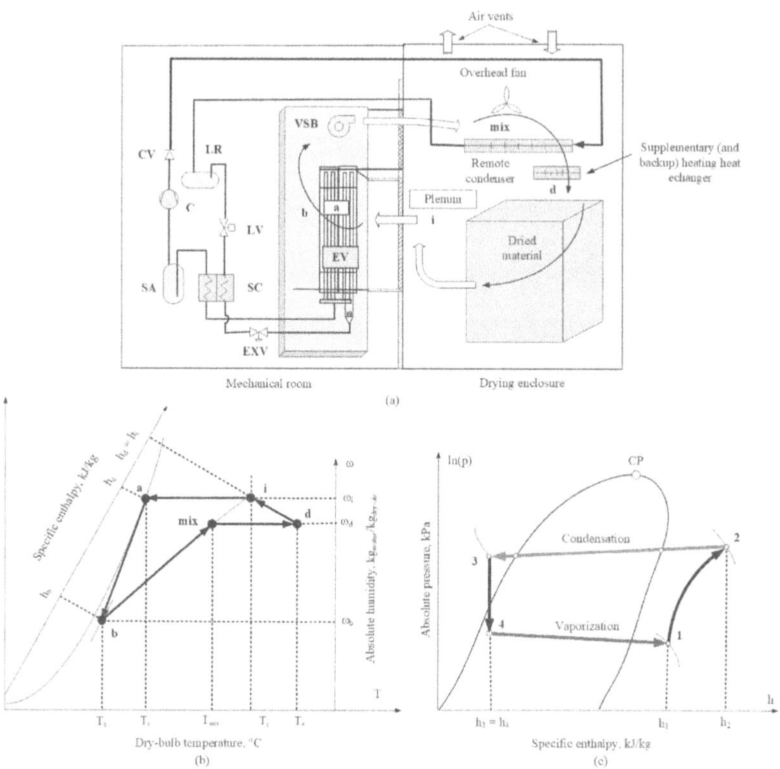

FIGURE 6.7 Batch drying chamber coupled with a *single-split* mechanical vapor compression heat pump. (a) Schematic. *Note:* Schematic not to scale; not all components shown. (b) Thermodynamic processes of heat pump refrigerant represented in ln(p)-h diagram. (c) Thermodynamic processes of drying air in the Mollier ($h-T$) diagram. C, compressor; CV, check valve; EV, evaporator; EXV, expansion valve; LV, refrigerant liquid valve; SA, suction accumulator; SC, subcooler; VSB, variable speed blower. *Note:* Schematic not to scale; not all components shown.

(process $4-1$), and becomes superheated vapor (state 1) (Figure 6.7c). It enters the compressor where is adiabatically compressed (process $1-2$) by the aid of electrical energy consumption. The compressed refrigerant superheated vapor at high pressure and high temperature (state 2) condenses inside the heat pump remote condenser and transfers heat to the drying air inside the batch dryer. After the condensing process, the subcooled refrigerant liquid at high pressure and high temperature (state 3) passes through an expansion valve where its pressure and temperature are lowered at constant specific enthalpy (process $3-4$) prior to entering the heat pump evaporator at state 4. Then, the heat pump thermodynamic cycle restarts.

Supplementary (and backup) heat exchangers supply, when necessary, some amounts of thermal energy to further heat the drying air after the heat pumps' remote condenser(s).

6.5.4 Coupling Batch Dryers with Multiple Parallel SPLIT Air-to-Air Mechanical Vapor Compression Heat Pumps

To fit the heat pumps' dehumidification (evaporators' thermal powers) capacities with materials' dewatering rates (which constantly decrease in time during the drying processes), several (multiple) air-to-air *split* mechanical vapor compression heat pumps can be installed in parallel (Figure 6.8). At the beginning of each drying

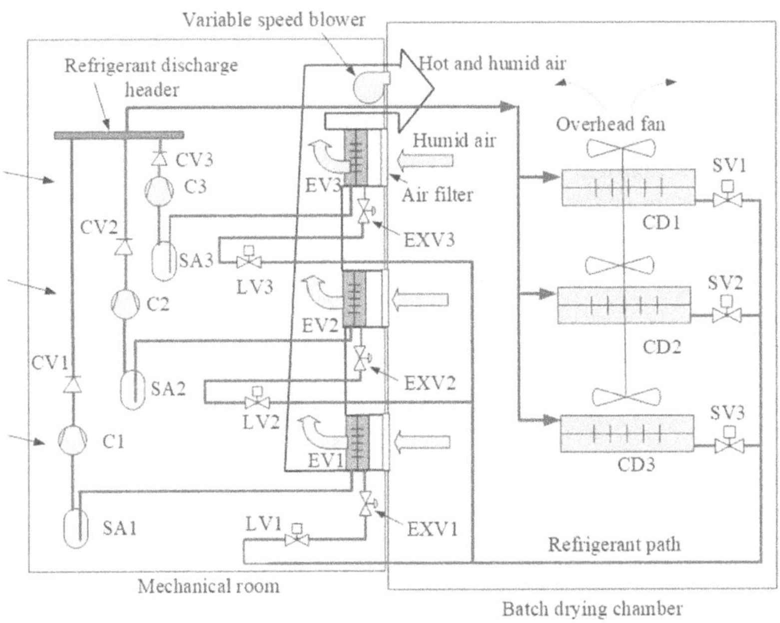

FIGURE 6.8 Schematic of a batch drying chamber coupled with *multiple* parallel *split* air-to-air mechanical vapor compression heat pumps. C, compressor; CV, check valve; EV1, low-temperature evaporator; EV2, high-temperature evaporator; EXV1, low-temperature expansion valve; EXV1, high-temperature expansion valve; LV, refrigerant liquid valve; SA, suction accumulator; SV, solenoid valve. *Note:* Schematic not to scale; not all components shown.

cycle, when the absolute humidity of drying air is at its maximum (after the dryer preheating period), all parallel mechanical vapor compression heat pumps should run. Progressively, as absolute humidity of drying air decreases, one by one the heat pumps shut down according to the appropriate preset drying values.

Such a coupling method allows matching the material dewatering rates with the heat pumps' dehumidification capacities by varying the number of heat pumps (compressors) running at the same time. Each running heat pump (compressor) uses a dedicated remote condenser (Minea 2011).

6.6 INDICATORS FOR DRYING PERFORMANCE

Heat pump-assisted dryers consist of two coupled thermodynamic systems: (i) air-to-air nonreversible mechanic vapor compression heat pumps; and (ii) hot air convective batch dryers.

Energy (drying) efficiency of each of these distinct thermodynamic systems can be defined as follows: (i) heat pump instantaneous coefficient of performance; (ii) heat pump overall coefficient of performance; and (iii) system (heat pump-assisted dryer) modified specific moisture extraction ratio.

6.6.1 Heat Pump Coefficient of Performance

Heat pump instantaneous coefficient of performance is defined as follows:

$$COP_{heat\ pump}^{instantaneous} = \dot{Q}_{cond} / \left(\dot{W}_{compressor} + \dot{W}_{blower} + \dot{W}_{controls} \right) \tag{6.25}$$

where

$COP_{heat\ pump}^{instantaneous}$ is the heat pump instantaneous coefficient of performance (–);
$\dot{Q}_{cond}$ is the condenser instantaneous thermal power (kW);
$\dot{W}_{compressor}$ is the compressor instantaneous electrical input power (kW);
$\dot{W}_{blower}$ is the drying air's blower instantaneous electrical input power (kW); and
$\dot{W}_{controls}$ is the controls instantaneous electrical input power (usually, negligible) (kW).

Heat pump overall coefficient of performance can be expressed as follows:

$$COP_{heat\ pump}^{overall} = Q_{cond}^{total} / \left[\left(\bar{\dot{W}}_{compressor} + \bar{\dot{W}}_{blower} \right) \cdot \tau_{cycle} \right] \tag{6.26}$$

where

$COP_{heat\ pump}^{overall}$ is the heat pump overall coefficient of performance (–);
Q_{cond}^{total} is the total thermal energy supplied by heat pump condenser to drying air during the entire drying cycle (kWh);
$\bar{\dot{W}}_{compressor}$ is the compressor average electrical input power over the entire drying cycle (kW);
$\bar{\dot{W}}_{blower}$ is the drying air's blower average electrical input power over the entire drying cycle (kW); and
τ_{cycle} is the duration of entire drying cycle (h).

6.6.2 System-Modified SMER and Modified SEC

Many published studies define the energy performances of heat pump-assisted dryers as total mass of extracted moisture (water) divided by total heat pump electrical energy consumed:

$$SMER_{heat\ pump} = m_{water\ extracted}^{total} / \left[\left(\dot{\bar{W}}_{compressor} + \dot{\bar{W}}_{blower}\right) \cdot \tau_{cycle}\right] \quad (6.27)$$

where

$SMER_{heat\ pump}$ is the heat pump–specific moisture extraction ratio (kg_{water} / kWh);
$m_{water\ extracted}^{total}$ is the total mass of moisture (water) extracted during the entire drying cycle (kg_{water});
$\dot{\bar{W}}_{compressor}$ is the compressor average electrical input power over the entire drying cycle (kW);
$\dot{\bar{W}}_{blower}$ is the drying air blower average electrical input power over the entire drying cycle (kW); and
τ_{cycle} is the duration of entire drying cycle (h).

However, a more appropriate energy performance indicator for heat pump-assisted dryers can be defined as follows:

$$MSMER_{system} = m_{water\ extracted}^{total} / \left[\left(\dot{\bar{W}}_{compressor} + \dot{\bar{W}}_{blower} + \dot{\bar{W}}_{dryer\ fas} + \dot{\bar{W}}_{dryer\ controls}\right) \cdot \tau_{cycle}\right] \quad (6.28)$$

where

$MSMER_{system}$ is the (drying system) modified specific moisture extraction ratio (kg_{water} / kWh);
$m_{water\ extracted}^{total}$ is the total mass of water extracted during the entire drying cycle (kg_{water});
$\dot{\bar{W}}_{compressor}$ is the heat pump compressor average electrical input power over the entire drying cycle (kW);
$\dot{\bar{W}}_{blower}$ is the heat pump blower average electrical input power over the entire drying cycle (kW);
$\dot{\bar{W}}_{dryer\ fans}$ is the dryer overhead fans average electrical input power over the entire drying cycle (kW);
$\dot{\bar{W}}_{controls}$ is the dryer and heat pump controls electrical input power over the entire drying cycle (*kW*); and
τ_{cycle} is the duration of entire drying cycle (h)

A modified specific energy consumption ratio could be defined as the reciprocal parameter of $MSMER_{system}$:

$$MSEC_{system} = \left[\left(\dot{\bar{W}}_{compressor} + \dot{\bar{W}}_{blower} + \dot{\bar{W}}_{dryer\ fas} + \dot{\bar{W}}_{controls}\right) \cdot \tau_{cycle}\right] / m_{water\ extracted}^{total} \quad (6.29)$$

where

$MSEC_{system}$ is the (drying system) modified specific energy consumption performance parameter (kWh/kg_{water});

$\dot{W}_{compressor}$ is the heat pump compressor average electrical input power over the entire drying cycle (kW);

$\dot{W}_{blower}$ is the heat pump blower average electrical input power over the entire drying cycle (kW);

$\dot{W}_{dryer\ fans}$ is the dryer overhead fans average electrical input power over the entire drying cycle (kW);

$\dot{W}_{controls}$ is the dryer and heat pump controls electrical input power over the entire drying cycle (kW);

τ_{cycle} is the duration of entire drying cycle (hours);

$m_{water\ extracted}^{total}$ is the total mass of moisture (water) extracted during the entire drying cycle (kg_{water}).

6.7 COMBINED (HYBRID) HEAT PUMP-ASSISTED DRYERS

Heat pump-assisted drying systems can be combined with other drying methods and energy sources (e.g., solar, infrared, radio frequency) to enhance the efficiency of moisture transfer mechanisms and drying rates (Figure 6.9).

6.7.1 Solar-Assisted Dryers

In geographical regions where the sun can supply thermal energy at temperatures higher than surrounding air temperatures, this energy can be utilized for auxiliary heating drying air either by direct heat transfer through solar radiation or by using stored solar energy. Thus, combined solar–heat pump-assisted dryers could further reduce energy consumptions and drying times.

Heat pump-assisted dryers combined with solar energy appears to be more efficient than heat pump-assisted drying and solar drying alone. According to some authors, in particular cases, the energy savings could reach 38% and the drying times could be reduced by up to 40% (Morrison 1994; Imre 2006; Sevik 2013).

A typical combined solar–heat pump-assisted dryer comprises a mechanical vapor compression heat pump (evaporator, condenser, compressor, and expansion valve) and a thermal solar collector. The solar collector acts as an additional air heater allowing further increasing the drying air temperature prior to entering the drying chamber (Figure 6.10a) or as the heat pump's direct expansion evaporator

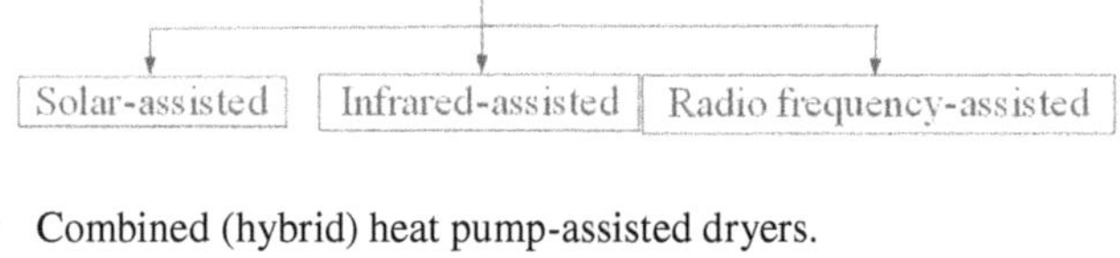

FIGURE 6.9 Combined (hybrid) heat pump-assisted dryers.

(Figure 6.10b). In solar-assisted heat pump dryers with indirect thermal solar collector (Figure 6.10a), the solar radiation heats the air passing through the solar collector. This heated air then serves to further heat, via an air-to-air heat exchanger, the drying air leaving the heat pump condenser, prior to being rejected outside. On the other hand, the hot and humid air leaving the drying chamber passes over the heat pump evaporator where it is cooled and dehumidified. Both sensible and latent heats are absorbed by the evaporating refrigerant and the resulting vapor is compressed by the compressor.

The high-pressure superheated refrigerant vapor condenses inside the condenser (*CD*) by transferring the recovered heat plus the compressor equivalent electrical energy consumed by the drying air. Ambient air is circulated through the thermal solar collector(s) by a variable speed fan, and then through an air-to-air heat exchanger where the drying air coming from the heat pump's condenser *CD* is further heated. After being further heated in this air-to-air heat exchanger, the hot and dry air passes through the drying chamber and picks up moisture from the dried product. Such a combined system offers the flexibility of operating with the solar collector only, with the heat pump only or, simultaneously, with both systems.

Such indirect solar-assisted heat pump dryers may provide the following advantages: (i) drying energy-savings; and (ii) higher operating drying temperatures compared to stand-alone heat pump drying systems. However, these systems have higher capital costs required for additional solar panels, blowers, heat exchangers, and controls, while the amount of available solar energy varies significantly throughout the day and/or the year.

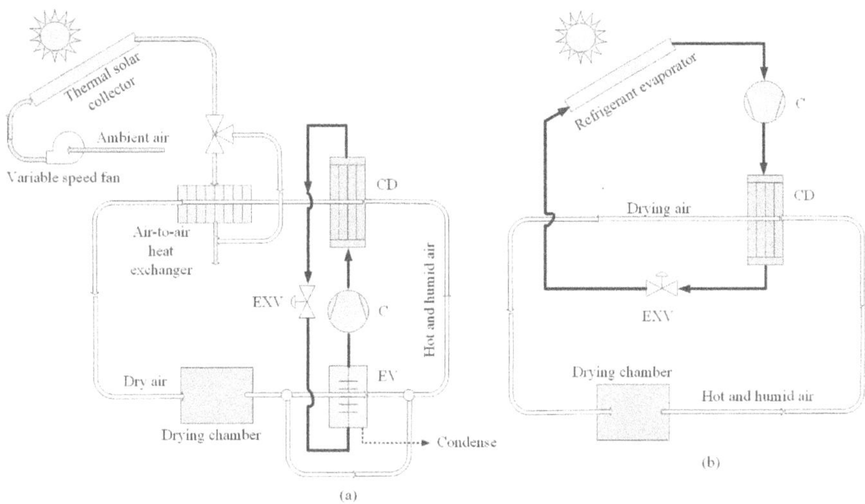

FIGURE 6.10 Schematics of combined solar–heat pump-assisted dryers. (a) With indirect thermal solar collector. (b) With direct expansion solar collector (refrigerant evaporator). *Note:* Schematics not to scale; not all components shown. *C*, compressor; *CD*, condenser; *EV*, evaporator; *EXV*, expansion valve.

Direct expansion solar-assisted heat pump dryers (Figure 6.10b) consist of thermal solar collectors acting as heat pumps' evaporators. In this case, the refrigerant is directly vaporized inside the solar collectors–evaporators due to the solar energy input (Hawlader et al. 2008).

Such concepts have advantages as the elimination of evaporators of traditional heat pump-assisted dryers and higher thermal efficiency of solar collectors. Although such a concept may reduce the system initial costs, the choice of the working fluid (refrigerant) according to the local ambient temperature conditions and the control strategy of the integrated hybrid system are very critical issues.

6.7.2 Radio Frequency–Assisted Dryers

Materials that are difficult to dry with hot air convective dryers because of their poor heat transfer characteristics can be good candidates for combined radio frequency–heat pump-assisted drying. Such hybrid dryers comprise mechanical vapor compression heat pumps coupled with pulsed, high-frequency radio frequency devices that generate volumetrically heat within the wet materials to heat them, and, simultaneously, vaporize the moisture at relatively low temperatures, usually not exceeding 82°C (Figure 6.11). The radio frequency generator generates heat within the wet material by the combined mechanism of dipole rotation and conduction effects (Thomas 1996; Marshall and Metaxas 1999).

In drying chambers, an alternating field is created between two large metal-plate electrodes which are connected to a high-frequency generator. The material is dried by placing or passing it between these electrodes. At radio frequency drying frequencies (1–30 MHz), the frictional heat is sufficient to produce temperatures exceeding

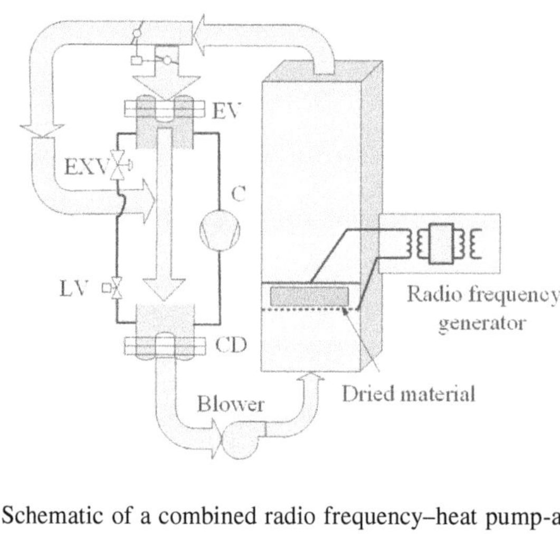

FIGURE 6.11 Schematic of a combined radio frequency–heat pump-assisted dryer. *Note:* Schematic not to scale; not all components shown. *C*, compressor; *CD*, condenser; *D*, air damper; *EX*, expansion valve; *EV*, evaporator; *F*, fan; *HR*, heat rejection coil; *M*, damper motor.

the boiling point of water. Such a drying process is rapid because the heat required to dry wet products is generated within the products themselves, and warping, surface discoloration, and cracking (caused by the stress of uneven shrinkage) associated with conventional drying methods are also avoided. However, this hybrid drying system may not be suitable for drying materials that have poor dielectric properties (Kudra and Mujumdar 2002).

Some characteristics of combined radio frequency–heat pump-assisted dryers (Kiang and Jon 2007) are as follows: (i) improves the color of the products, especially those that are highly susceptible to surface color change since radio frequency drying starts from the internal to the product surface; and (ii) materials cracks, caused by the stresses of uneven shrinkage, can be eliminated.

6.7.3 Infrared–Assisted Dryers

Infrared heaters can be used as auxiliary energy sources combined with heat pump-assisted dryers (Figure 6.12). Infrared light radiation can raise the temperature of dried materials without heating up the drying air. It is an energy-intensive process that rapidly rises the temperatures of dried materials.

Combined infrared–heat pump-assisted dryers could be used for fast removal of surface moisture during the initial stages of drying, followed by intermittent drying over the rest of the drying process. Heat for drying is generated by radiation from infrared generators. Incorporating infrared into an existing heat pump-assisted dryer is simple and the capital cost is low. This mode of operation ensures a faster initial drying rate and offers advantages of compactness, simplicity, ease to control, and low equipment cost. There are possibilities of significant energy-savings and enhanced product quality due to reduced residence time in the dryer chamber (Jones 1992).

Some advantages of combined infrared–heat pump-assisted dryers are as follows (Paakkonen et al. 1999; Zbicinski et al. 1992): (i) ensure high heat transfer

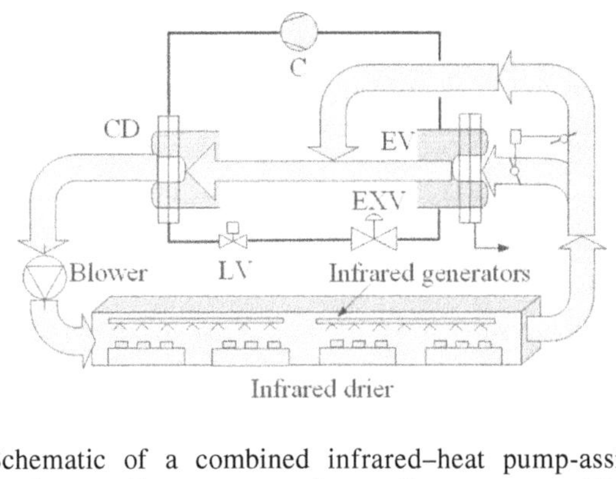

FIGURE 6.12 Schematic of a combined infrared–heat pump-assisted dryer. *Note:* Schematic not to scale; not all components shown. *C*, compressor; *CD*, condenser; *D*, air damper; *EX*, expansion valve; *EV*, evaporator; *F*, fan; *HR*, heat rejection coil; *M*, damper motor.

rates (up to 100 kW/m^2); (ii) easy to direct the heat source to the product surface; (iii) in the case of food, heat-sensitive products can reduce the drying times; (iv) quick response times allowing easy control of drying processes; (v) may provide faster initial drying rate due to the direct transfer of energy from the infrared heating elements to product surfaces without heating the surrounding air; (vi) system compactness, simplicity, and relatively low cost; and (vii) possibility of significant energy-savings and enhanced product quality due to reduced residence time in drying chambers.

6.8 ENVIRONMENTAL ASPECTS CONCERNING HEAT PUMP–ASSISTED WOOD DRYERS

One of the environmental issues during the drying process of woods with conventional convective hot air dryers consists of emissions of volatile organic compounds (VOCs) by the exhausted drying air. The rate of VOCs emissions depends on the temperature, species (more common for softwoods than for hardwoods), amount of water loss, and presence of knots, which can contain even more than ten times higher VOCs than the normal wood (Table 6.4). VOCs emissions increase with the increasing dry-bulb temperature of drying air, especially due to the thermal degradation of wood components.

In the case of heat-pump-assisted wood drying, which is performed with practically closed batch dryers, the liquid moisture condensate contains toxic VOCs such as formaldehyde and acetaldehyde that are harmful to trout and *Daphnia*. The wood drying companies that are not connected to municipal sewer systems have to use water treatment technologies to reduce the concentration of toxic substances to levels below the limits set by the local or national regulations or must consider some fees for collecting and transporting water to the municipal waste disposal site. If the dryers' sites are connected to sewer systems, the municipalities may accept the (liquid) condensate but charge to intercept and treat the effluents.

Dryers with 354 m^3 of softwood capacities are average industrial-scale dryers in Canada. The volume of water extracted from softwoods with high-temperature heat pumps in such dryers is less than 15 m^3/day. But the water extracted is acid (with pH generally between 3.3 and 4.3) and contains low concentrations of organic loads,

TABLE 6.4
VOC Emissions

Wood Species	Dry-Bulb Temperature (°C)	Lumber Board Thickness (mm)	VOC (mg/kg$_{oven\text{-}dry\ wood}$)
Southern pine	82	50	891
	118	50	1,110
Radiata pine	120	35	200
	140	35	400
White fir	82	50	49
	115	50	101

TABLE 6.5
Problematic Pollutants in Water Extracted from Dried Softwoods

		Protection of Aquatic Life	
Parameter	**Concentration (mg/L)**	**Acute Toxicity Limit (μg/L)**	**Chronic Effect Limit (μg/L)**
Formaldehyde	34–5,700	1,000	120
Acetaldehyde	200–1,500	1,200	–

such as total biochemical oxygen demand (BOD) (<320 mg/L) and total chemical oxygen demand (COD) (<454 mg/L). Problematic pollutants are the formaldehyde (CH_2O) (which can exceed the acute toxicity limit by up to 5.7 times), and to a lesser degree, the acetaldehyde (CH_3CHO) (which can exceed the acute toxicity limit by up to 1.3 times) (Table 6.5). These substances are toxic VOCs being harmful to trout and *Daphnia* (Minea 2008).

Formaldehyde is an organic molecule normally found in the air at concentrations lower than 1 $\mu g/m^3$ and in rainwater at concentrations approaching 10 μg/L. It quickly becomes toxic at higher concentrations. Formaldehyde is biodegradable and its molecules degrade through photolysis. In the air, it has a half-life of 4.1 hours when submitted to direct solar radiation. When dissolved in water under natural conditions, it is not volatile and forms methylene glycol. Since formaldehyde and acetaldehyde both are fully biodegradable, it is assumed that their removal rate is proportional to the *BOD* removal rate. The condensate treatment options may be selected taking into consideration factors such as the capacity to reduce the organic contaminant content in water in low concentrations (effluent *COD* concentration of less than 500 ppm) and the availability and ability of equipment for treating wastewater volumes below $15\,m^3$/day. Consequently, the wood drying companies that are not connected to municipal sewer systems have to use water treatment technologies to reduce the concentration of toxic substances to levels below the limits set by the Canadian regulations. If the kiln site is connected to a sewer system, the municipality may accept the condensate, but charge to intercept and treat the effluent. If the kiln is not served by a local sewer system, fees for collecting and transporting water to the municipal waste disposal site must be considered.

The condensate treatment options may be selected taking into consideration factors such as the capacity to reduce the effluent *COD* concentration of less than 500 ppm, and the availability and ability of equipment for treating wastewater volumes below15 m^3/day.

Options to consider for condensate treatment are the activated carbon adsorption, natural mitigation, biofiltration, and reverse osmosis. Advanced oxidation techniques, such as β-radiation, UV-catalyzed chemical oxidation, ozonation, and UV photolysis, can be added to this list.

To adequately treat the (liquid) condensates with as low as possible investments, operation costs, and electricity energy consumptions, natural mitigation and ultraviolet (UV) peroxydation technologies are among the most efficient (Minea 2008).

TABLE 6.6
Summary of Final Condensate Concentrations

	Final Concentration					
Technology	**COD (mg/L)**		**BOD (mg/L)**		**CH_2O (µg/L)**	**CH_3CHO (µg/L)**
Natural mitigation	150	67	16	95	<285	<75
UV peroxidation	91	80	36	89	<647	<170

Notes: BOD, biochemical oxygen demand; CH_2O, formaldehyde, COD, chemical oxygen demand; UV, ultraviolet.

Analysis based on the capacity to attain standard norms, low investment and operating costs, and quick start-up capability shows that the optimum technology would be the UV peroxidation, followed by natural mitigation. Table 6.6 gives the final concentrations and percentages of reduction for both recommended treatment technologies.

In terms of water treatment performance, the *UV* peroxidation process is similar to ozonation, but it offers the advantage of significantly lower investment and operating costs because it uses a stable oxidant manufactured at lower cost in a factory. As with ozonation, the process can be easily restarted after long shutdowns, and the peroxide decomposes quickly without leaving residues. The natural mitigation is one of the most inexpensive processes in terms of investment and operating costs. Purification is performed over a period of 70 days in a series of several watertight basins. In general, three basins are used: The first one reduces the organic load as much as possible, while the other two act as buffer zones if the first basin overflows. The purification mechanisms combine solar UV radiation, input of atmospheric oxygen and surface algae photosynthesis, aerobic digestion at the surface, and anaerobic digestion on the bed of the basin. More than 85% of the *COD* can be eliminated from the liquid phase at the lagoon outflow, a part of which is left in the sludge that settles on the bed of the basin. This technology is however less effective in winter and requires long treatment periods and significant surface areas, up to $1,050\,m^3$ to treat $15\,m^3$/day of condensate effluent. Natural mitigation methods make use of resistant microorganisms, but they remain subject to toxic shocks and temperature variations, and restarting can be a lengthy process.

REFERENCES

Carrington, C.G., P. Bannister, Q. Liu. 1995. Performance analysis of a dehumidifier using HFC-134a. *International Journal of Refrigeration* 18:477–488.

Cech, M.J., F. Pfaff. 1993. *Operator Wood Drier Handbook for East of Canada*, FORINTEK Corp., Canada's Eastern Forester Products Laboratory, Vancouver, Canada, 189 pp.

Chou, S.K., K.J. Chua. 2007. *Heat pump drying systems*. In: *Handbook of Industrial Drying*, 3rd edition, edited by A.S. Mujumdar, Taylor & Francis Group, Abingdon, UK.

Chua, K.J., S.K. Chou, A.S. Mujumdar, J.E. Ho, M.N.A. Hawlader. 2002. Heat pump drying: Recent developments and future trends. *Drying Technology: An International Journal*, 20(8):1559–1577.

Hawlader, M.N.A., S.M.A. Rahman, K.A. Jahangeer. 2008. Performance of evaporator–collector and air collector in solar assisted heat pump dryer. *Energy Conversion and Management* 49:1612–1619.

Imre, L. 2006. Solar drying. In: *Handbook of Industrial Drying*, edited by A.S. Mujumdar, Taylor & Francis Inc., Boca Raton, FL, pp. 317–319.

Jones, P. 1992. Electromagnetic wave energy in drying processes. In: Drying 92, edited by A.S. Mujumdar, Elsevier Science Publisher B.V., pp. 114–136.

Kiang, C.S., C.K. Jon. 2007. Heat pump drying systems. In: *Handbook of Industrial Drying*, edited by A.S. Mujumdar, Taylor & Francis Inc., Boca Raton, FL, pp. 1104–1105.

Kudra, T., A.S. Mujumdar. 2002. *Advanced Drying Technologies*, Marcel Dekker, New York, 457 pp.

Léger, F., M. Amazouz. 2003. Évaluation des pratiques de control et de commande des séchoirs à bois (in French). Centre de la technologie de l'énergie de CANMET, Québec, Canada.

Lewis, D. 2003. High temperature dehumidification systems. U.S. Patent No. 20030208923.

Marshall, M.G., A.C. Metaxas. 1999. Radio frequency assisted heat pump drying of crushed brick. *Applied Thermal Engineering* 19(4):375–388.

McQuiston, F.C., J.D. Parker. 1988. Heating, Ventilating, and Air Conditioning: Analysis and Design, 3rd edition, John Wiley & Sons, Inc., New York, NY.

Minea, V. 2004. Heat pumps for wood drying: New developments and preliminary results. In: *Proceedings of the 14th International Drying Symposium (IDS 2004)*, Volume B, pp. 892–899. Sao Paulo, Brazil, August 22–25.

Minea, V. 2006. Hardwood drying with low temperature heat pumps. In: *Proceedings of the 15th International Drying Symposium*, Volume C, pp. 1757–1762. Budapest, Hungary, August 20–23.

Minea, V. 2008. Energetic and ecological aspects of softwood drying with high-temperature heat pumps. *Drying Technology* 26(11):1373–1381.

Minea, V. 2010a. Improvements of high-temperature drying heat pumps. In: *Proceedings of the International Refrigeration and Air Conditioning Conference at Purdue*, Purdue, IN, July 12–15.

Minea V. 2010b. Improvements of high-temperature drying heat pumps. *International Journal of Refrigeration* 33(1):180–195.

Minea, V. 2011. *Industrial drying heat pumps.* In: *Mechanical Engineering Theory and Applications: Refrigeration—Theory, Technology and Applications*, edited by Mikkel E. Larson, Nova Science Publishers Inc., New York, NY, pp. 1–70.

Minea, V. 2012. Efficient energy recovery with wood drying heat pumps. *Drying Technology* 30:1630–1643.

Minea, V. 2019. *Industrial Heat Pump-Assisted Wood Drying*, CRC Press, Boca Raton, FL.

Morrison, G.L. 1994. Simulation of packaged solar heat-pump water heaters. *Solar Energy* 53(3):249–257.

Mujumdar, A.S., L.X. Huang. 2007 Global R&D needs in drying. *Drying Technology* 25:647–658.

Paakkonen, K., J. Havento, B. Galambosi, M. Pyykkonen. 1999. Infrared drying of herbs. *Agricultural and Food Science in Finland* 8:19–27.

Radermacher, R., Y. Hwang. 2000. *Vapor Compression Heat Pumps with Refrigerant Mixtures*, Center for Environmental Energy Engineering, Department of Mechanical Engineering, University of Maryland, College Park, Baltimore, MD.

Sevik, S. 2013. Design, experimental investigation and analysis of a solar drying system. *Energy Conversion and Management* 68(1):227–234.
Strumillo, C. 2006. Perspectives on development in drying. *Drying Technology* 24:1059–1068.
Thomas, W.J. 1996. RF drying provides process savings: New systems optimize radio frequency drying for the ceramic and glass fiber industries. *Ceramic Industry Magazine*, 30–34.
Zbicinski, I., A. Jakobsen, J.L. Driscoll. 1992. Application of infrared radiation for drying of particulate material. In: *Drying 92*, edited by A.S. Mujumdar, Elsevier Science Publisher B.V., pp. 704–711.

7 Optimizing Dehumidification Capacity of Mechanical Vapor Compression Heat Pumps

7.1 INTRODUCTION

Drying is a complex thermodynamic process that depends on the dried material type, physical/chemical structure, and its initial average moisture content and temperature.

The role of air-to-air mechanical vapor compression heat pump in heat pump-assisted batch dryers is to recover latent and sensible heat from the moist and warm drying air that passes over the external heat transfer surfaces of heat pumps' evaporators. This recovered thermal energy, plus the equivalent heat of the electrical energy consumed by the heat pumps' compressors and blowers, serves to overcome the water retention forces inside the pieces of dried materials and to remove moisture from their surfaces. These simultaneous thermodynamic processes must be carefully balanced to achieve thermally equilibrated drying processes.

As the moisture content drops at surfaces of wet porous solids, the water migrates from inside of dried pieces to their surfaces mainly because of moisture content differences. The migration rate of moisture depends on the liquid water diffusivity, which is a function of local moisture content and temperature. For cost-effective operation of heat pump-assisted dryers, the migration rate of water to the materials' surfaces should match the moisture removal rates from materials' surfaces. In other words, to ensure heat pump-assisted drying systems to operate practically without cycling the heat pumps' compressors, the moisture removal rates and heat pumps' (nominal) dehumidification capacities must be, as far as possible, continuously matched. This difficult but fundamental task may be achieved by sizing the heat pump nominal dehumidification capacity (i.e., evaporator nominal thermal power) according to the dryer capacity (volume) and the product initial mass and its initial and final moisture contents. If the heat pump nominal dehumidification capacity is oversized, equivalent heat corresponding to the electrical energy consumed (compressor and blower) will progressively build up inside the heat pump thermodynamic cycle, and, thus, it must be periodically rejected outdoor by means of external parallel air- or water-cooled condenser(s), desuperheater(s), or refrigerant subcooler(s). Such heat rejection means should be avoided as much as possible to optimize the heat pump-assisted dryer energy performance.

DOI: 10.1201/9781003618386-7

The ratio between mass of moisture removed by the heat pump-assisted dryer (expressed in kg_{water}) and the system total electrical energy consumed by heat compressor, blower and controls, and dryer overhead fans and controls (expressed in kWh; called modified specific moisture extraction ratio [MSMER]) should be as high as possible. This means that adequate (sufficient) quantity of dried material (and initial total moisture content) should be supplied to the drying enclosure to provide enough moisture, and thus enough (latent) thermal input to the heat pump evaporator.

If the evaporator thermal dehumidification capacity is not matched with the material dewatering rate during each period of drying cycles, the heat pumps' coefficients of performance ($COP_{heat\ pump}$) and dryers' MSMER may not be accurately determined. In other words, appropriate correlation between the initial mass and moisture content, the moisture extraction rates from the dried products, and the heat pump dehumidification capacities have to be ensured during each period of drying cycles.

Insufficient initial quantities of dried material may provide (i) small quantities of moisture and latent heat extraction, as well as too low evaporating and compressor suction pressures and temperatures; (ii) risks of humidity freezing on the evaporator external heat transfer finned surface; and (iii) frequent and/or too short compressor on/off cycles. Such (undesirable) poor operating conditions may lead to heat pump operating troubleshooting, reduce the compressor technical life, and falsify the system overall energy performances.

7.2 METHODS TO MATCH THERMAL CAPACITIES

The mass of removed (condensed) moisture strongly depends on the drying air mass flow rate (and velocity) across the evaporator external heat transfer surface. Because the evaporator heat exchange surface is constant, it is necessary to vary the drying air mass flow rate (and velocity) to provide quasi-constant dehumidification rates that should match with materials' thermal dewatering capacities.

The complexity of the control of coupled hot air convective batch dryers with air-to-air mechanical vapor compression heat pumps comes from the requirement to match the material thermal dewatering rate (i.e., the material surface moisture evaporation rate) with the heat pump dehumidification capacity (i.e., evaporator thermal power) throughout the entire drying cycle. It is one of the most critical issues for drying processes with heat pumps. This requirement becomes mandatory mainly because toward the end of the drying cycles, the material thermal dewatering rates progressively decrease. Therefore, the heat pump dehumidification capacity must also progressively decrease to improve the overall energy efficiency of the entire drying process without stopping many times the heat pump compressor. In this way, the compressors of air-to-air mechanical vapor compression heat pumps will be able to run quasi-continuously to produce consistent product throughputs.

In addition, the temperature difference between the evaporating and condensing temperatures for single-stage air-to-air mechanical vapor compression heat pumps should not exceed 40°C to maintain the heat pump average coefficients of performance in economically efficient ranges (usually, between 3 and 4), and to minimize the amount of total electrical energy consumed.

The following are the optimum strategies to match the materials' dewatering capacities with heat pumps' actual dehumidification rates: (i) varying the mass flow rate of drying air across the evaporator heat transfer surface; (ii) varying the velocity of drying air across the evaporator heat transfer surface; (iii) using intermittent drying strategy; and (iv) using multiple parallel heat pump compressors.

7.2.1 Batch Dryers Coupled with Compact Heat Pumps: Variable Mass Flow Rate of Drying Air over Evaporator, and Constant Speed Blower

Because the moisture content of dried products is relatively high at the beginning of drying processes (dried materials being able to provide high absolute humidity to the drying air), and much lower at the end, it is necessary to continuously decrease the drying air flow rate of the warm and humid air passing over the heat pump's evaporator external heat transfer surface. This operation allows at any time to sufficiently cool the warm and humid drying air to reach its dew point, and then to provide almost constant latent heat (i.e., evaporator dehumidification thermal power).

Continuously varying the drying air flow rate over the evaporator and thus over the condenser heat transfer surfaces is mandatory owing to the following reasons (Minea 2015):

a. The material thermal dewatering rate is relatively high at the beginning and relatively low toward the end of drying cycles; therefore, the heat pump dehumidification capacity must be continuously matched with the material thermal dewatering rate.
b. If the drying air flow rate through the evaporator is constant and too high, the quantity of moisture extracted (condensed) will be proportionally low and the heat extracted (by moisture condensing) will be insufficient for an efficient drying process.
c. If the total (sensible and latent) heat input at the heat pump evaporator inlet is constant and too low, the compressor may trip off on its low suction pressure control.
d. If the drying air flow rate over the condenser is constant and too high, the compressor may shutdown on its low limit pressure.
e. If the drying air flow rate over the condenser is constant and too low, the compressor may trip off on its high pressure.

A relatively efficient method to control the heat pump dehumidification rate (i.e., the heat pump's evaporator actual thermal power) and the drying air flow rates over the evaporator and condenser consists of using coil face-and-bypass dampers and constant speed air blowers which allow varying the drying air flow rate (and velocity) over the evaporator finned surface by partially bypassing it (Lewis 2003; Minea 2006) (Figure 7.1).

In this case, drying air and refrigerant temperature sensors and motorized dampers (*MD*) control both primary and bypass drying air flow rates so that the temperature of the drying air leaving the evaporator be approximately equal to the refrigerant

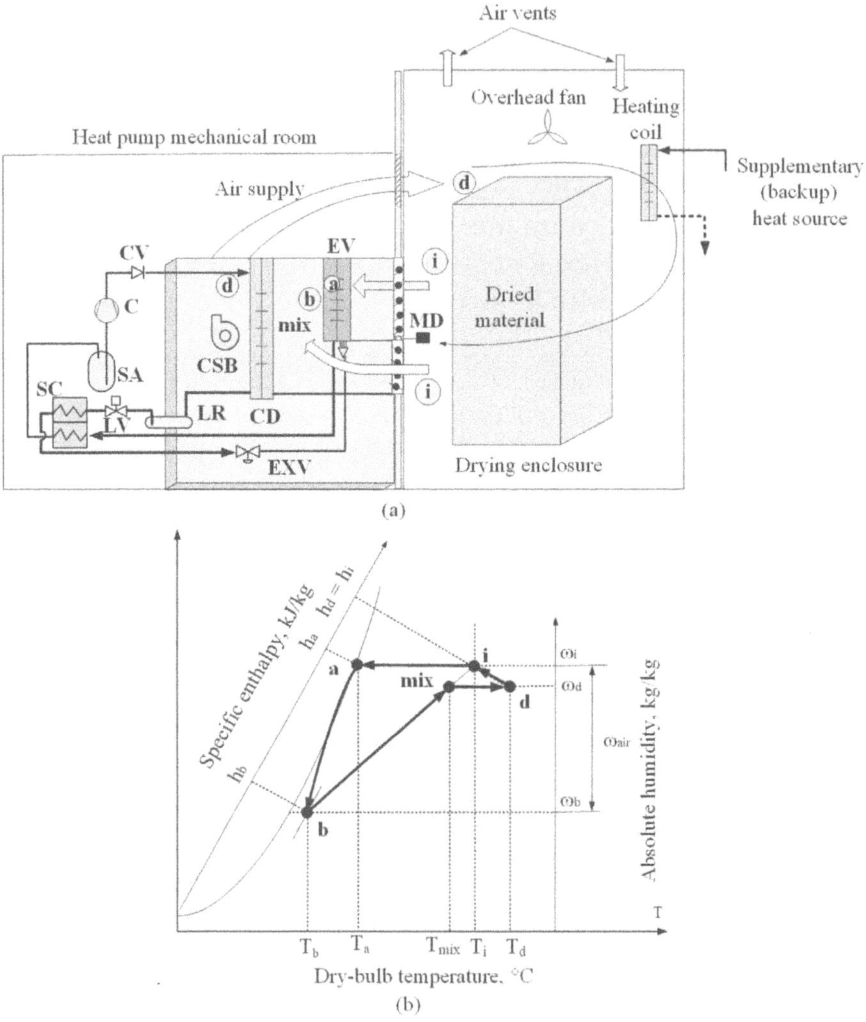

FIGURE 7.1 Batch dryer coupled with *compact* mechanical vapor compression heat pump: Variable mass flow rate of drying air over evaporator and constant speed blower. (a) Schematic. *Note:* Not to scale; not all components shown. (b) Thermodynamic processes of drying air represented in the Mollier ($h-T$) diagram. *C*, compressor; *CD*, heat pump remote condenser; *CSB*, constant speed drying air blower; *EV*, evaporator; *EXV*, expansion valve; *LV*, refrigerant liquid valve; *MD*, motorized dampers.

temperature leaving the heat pump evaporator. For industrial-scale applications, it is however a relatively imprecise method because of the low reading accuracy of temperature sensors and/or poor reliability of motorized dampers.

By neglecting the recovered sensible heat, the evaporator (latent) thermal power (dehumidification capacity) can be expressed as follows (Figure 7.1b):

$$\dot{Q}_{evap} = \dot{m}_{water} \cdot h_{water}^{latent} = \dot{m}_{air} \cdot \Delta\omega_{air} \cdot h_{water}^{latent} \tag{7.1}$$

where

$\dot{Q}_{evap}$ is the evaporator dehumidification (thermal power) capacity (kW);
$\dot{m}_{water}$ is the mass flow rate of condensed moisture (kg_{water}/s);
h_{water}^{latent} is the water condensing enthalpy (latent heat) (kJ/kg_{water});
$\dot{m}_{air}$ is the mass flow rate of the drying air flowing over the evaporator (kg_{air}/s);
$\Delta\omega_{air} = \omega_i - \omega_b$ is the variation of absolute humidity of drying air flowing over the evaporator (kg_{water}/kg_{air});
ω_i is the drying air absolute humidity at the evaporator inlet ($kg_{water}/kg_{dry\,air}$); and
ω_b is the drying air absolute humidity at the evaporator outlet ($kg_{water}/kg_{dry\,air}$).

If the objective is to keep $\dot{Q}_{evap}$ constant as long as possible during the entire drying cycle, i.e., to remove (condensate) approximately constant quantities (masses) of moisture ($\Delta\omega_{air}$) during the entire drying cycle, the mass flow rate of drying air over the evaporator finned heat transfer surface should be varied according to the following equation:

$$\dot{m}_{air} = \dot{Q}_{evap} / \left(\Delta\omega_{air} \cdot h_{water}^{latent}\right) \simeq Const\,1 / \left(\Delta\omega_{air} \cdot h_{water}^{latent}\right) \tag{7.2}$$

where

$\dot{m}_{air}$ is the mass flow rate of drying air over the evaporator (kg_{air}/s);
$\dot{Q}_{evap}$ is the evaporator dehumidification (thermal power) capacity (kW);
$\Delta\omega_{air}$ is the variation of drying air absolute humidity over the evaporator heat transfer surface ($kg_{water}/kg_{dry\,air}$);
h_{water}^{latent} is the water condensing enthalpy (latent heat) (kJ/kg_{water}); and
$Const\ 1 \simeq \dot{Q}_{evap}$.

From Equation 7.2, it can be seen that

$$\text{(a) if } \dot{m}_{air} \text{ is low} \left(\text{or decreases}\right), \Delta\omega_{air} \text{ is high} \left(\text{or increases}\right) \tag{7.3}$$

$$\text{(b) if } \dot{m} \text{ is high} \left(\text{or increases}\right), \Delta\omega_{air} \text{ is low} \left(\text{or decreases}\right) \tag{7.4}$$

At the beginning of drying cycle, the dried material is very humid (wet), and thus the absolute humidity of the drying air is at its highest (maximum) value. As a result, the drying air absolute humidity variation ($\Delta\omega_{air}$) over the evaporator external heat exchange surface is also high. For that, the mass flow rate of the drying air over the evaporator external heat transfer surface must be relatively low to extract as much as possible moisture from the drying air. In other words, lowering the drying air flow rate over the evaporator heat transfer area helps to increase the moisture condensation rate.

On the other hand, toward the end drying cycle, the moisture content of the dried product, as well as the absolute humidity of the drying air at the evaporator inlet, decreases progressively. Consequently, the drying air flow rate over the evaporator external heat transfer surface should be progressively increased to lower the quantity of moisture removed (condensed), without stopping the heat pump compressor.

Such a control of the drying air flow rate over the heat pump evaporator finned heat transfer surface aim to keep the heat pump actual dehumidification capacity (evaporator thermal power) as constant as possible and, at the same time, to efficiently match it with the material actual thermal dewatering rate.

7.2.2 Batch Dryers Coupled with Split Heat Pumps: Variable Velocity of Drying Air over Evaporator, with Variable Speed Blower (VSB)

In the case of hot air convective batch dryers coupled with *split* mechanical vapor compression heat pumps, the material dewatering rate could be matched with the evaporator thermal power (heat pump dehumidification capacity) by varying the velocity (and mass flow rate) of drying air over the evaporator external heat transfer surface using VSBs (Figure 7.2).

Equation 7.1 gives:

$$\dot{m}_{air} = \overline{\rho}_{air} \cdot A_{evap} \cdot \dot{w}_{air} = \dot{Q}_{evap} / \left(\Delta\omega_{air} \cdot h_{water}^{latent}\right) \simeq Const\ 1 / \left(\Delta\omega_{air} \cdot h_{water}^{latent}\right) \quad (7.5)$$

where

$\dot{m}_{air}$ is the mass flow rate of drying air over the evaporator (kg_{air}/s);
$\overline{\rho}_{air}$ is the average mass density of drying air (kg_{air}/m^3);
A_{evap} is the evaporator external heat transfer area (m^2);
$\dot{w}_{air}$ is the drying air velocity over the evaporator finned surface (m/s);
$\dot{Q}_{evap}$ is the evaporator dehumidification (thermal power) capacity (kW);
$Const\,1 \simeq \dot{Q}_{evap}$ is the value of the evaporator thermal power (dehumidification capacity) intended to be as much as possible constant over the entire drying cycle (kW);
$\Delta\omega_{air}$ is the variation of drying air absolute humidity over the evaporator heat transfer surface ($kg_{water}/kg_{dry\,air}$); and
h_{water}^{latent} is the water condensing enthalpy (latent heat) (kJ/kg_{water}).

From Equation 7.5 results the velocity of drying air over the evaporator external finned heat transfer surface:

$$\dot{w}_{air} = \left[\dot{Q}_{evap} / \left(\Delta\omega_{air} \cdot h_{water}^{latent} \cdot \overline{\rho}_{air} \cdot A_{evap}\right)\right] = Const\ 1 / \left(\Delta\omega_{air} \cdot Const\ 2\right) \quad (7.6)$$

where

$Const\,2 = h_{water}^{latent} \cdot \overline{\rho}_{air} \cdot A_{evap}$ is a second constant

From Equation 7.6, it can be seen that

$$\text{(a) if } \dot{w}_{air} \text{ is high (or increase)}, \Delta\omega_{air} \text{ is low (or decreases)} \quad (7.7)$$

$$\text{(b) if } \dot{w}_{air} \text{ is low (or decreases)}, \Delta\omega_{air} \text{ is high (or increases)} \quad (7.8)$$

Equations 7.7 and 7.8 show that to keep $\dot{Q}_{evap}$ approximately constant ($\dot{Q}_{evap} \simeq Const\ 1$) during the entire drying cycle (i.e., from the beginning to the end), another possibility

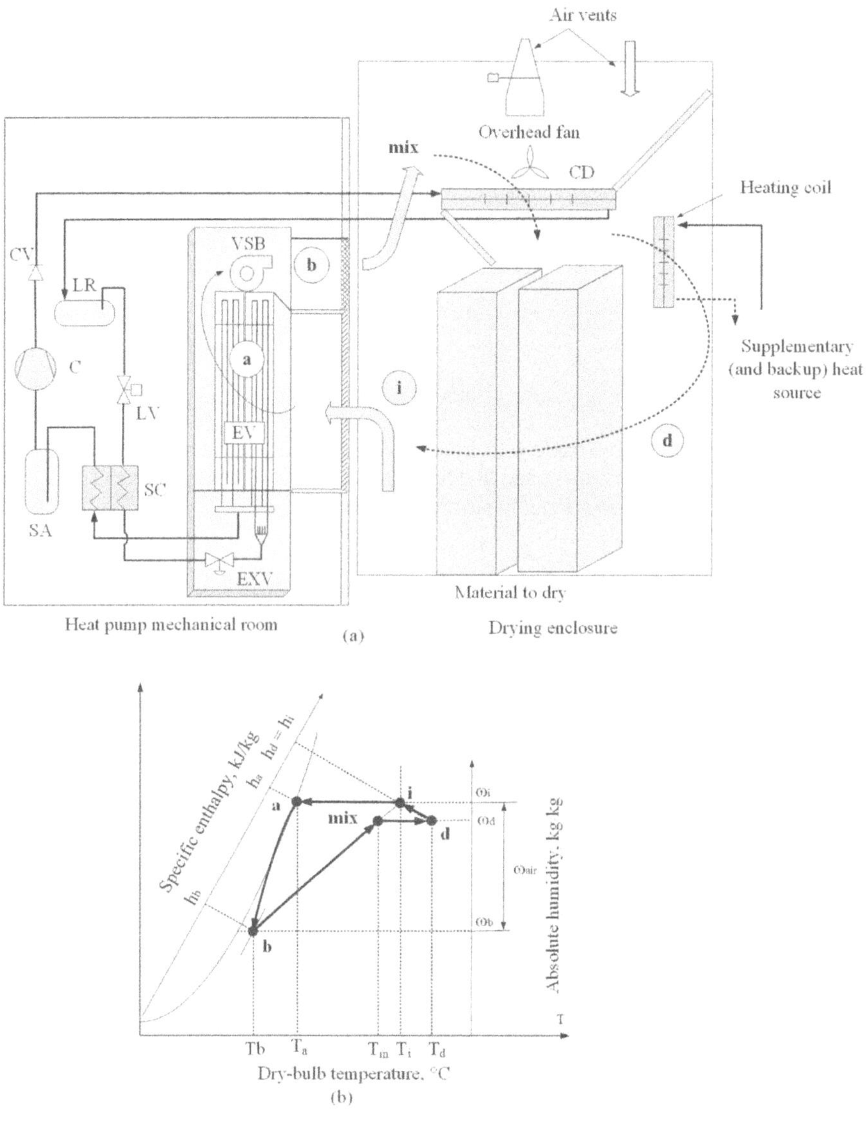

FIGURE 7.2 Batch dryer coupled with *split* mechanical vapor compression heat pump—variable velocity of drying air over evaporator with VSB. (a) Schematic. *Note:* Not to scale; not all components shown. (b) Thermodynamic processes of drying air represented in the Mollier ($h-T$) diagram; *B*, variable speed drying air blower; *C*, compressor; *CD*, heat pump remote condenser; *EV*, evaporator; *EXV*, expansion valve; *LV*, refrigerant liquid valve; *SC*, refrigerant subcooler; *VSB*, variable speed blower.

is to *vary* the drying air velocity ($\dot{w}_{air}$) over the evaporator external heat transfer surface by using VSBs. If the air velocity ($\dot{w}_{air}$) is *low* at the beginning of drying cycle, $\Delta\omega_{air}$ will be high. On the contrary, if $\dot{w}_{air}$ is high at the end of drying cycle (when the moisture content of the dried material progressively decreases), $\Delta\omega_{air}$ will be

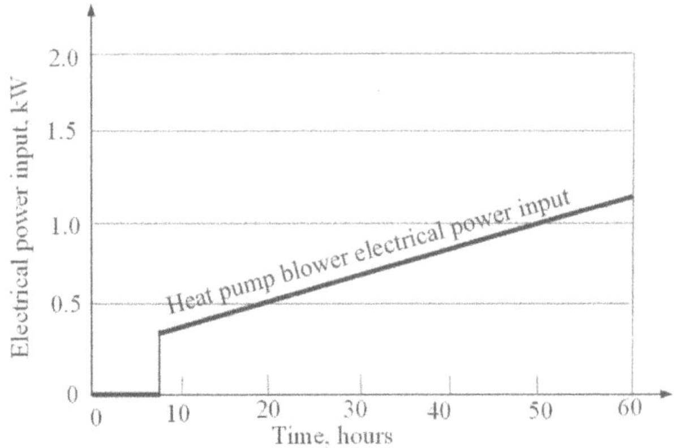

FIGURE 7.3 Example of variable velocity of drying air over the evaporator external heat transfer surface. (a) Low, medium, and high increase of blower electrical current frequency. (b) Increase of electrical power input to heat pump VSB.

low. Thus, the variation of absolute humidity ($\Delta\omega_{air}$) of drying air over the evaporator external heat exchange surface will proportionally decrease from the beginning through the end of drying cycle.

Figure 7.3 shows, as a typical example, that by progressively increasing the frequency of the electrical current supplying the heat pump VSB from 25 Hz to 60 Hz, the blower electrical power input increased from 0.3 kW (corresponding to relatively high moisture removal/dehumidification at the beginning of drying cycle) to 1.2 kW (corresponding to relatively low moisture removal/dehumidification rates toward the end of drying cycle) (Minea 2019).

7.2.3 Batch Dryers Coupled with Compact Heat Pumps: Intermittent Drying with Evaporator Air Bypass and Constant Speed Blower

Another technique that allow matching the heat pump dehumidification capacity (evaporator thermal power) with the material dewatering rate consists of intermittent operation of heat pump compressor (Figure 7.4). Such a method is well-adapted to capillary porous materials (as woods) because their drying processes are often accompanied by phenomena as shrinkage, when, after a certain time after the beginning of drying cycle, water on the material surface disappears. It consists of periodically stopping the heat pump compressor for certain (predetermined) periods of time. During the compressor shutdown (predetermined) periods, the moisture (water) movement from the material interior to its surface allows remaking the water liquid film. After the compressor restart, the quantity of liquid water that will be removed from the material surface will be again sufficiently high, the material cracking avoided, the moisture extraction rate optimized, and the final quality of dried materials (as woods) preserved. In other words, during the tempering (compressor stopping) periods, the capillary transportation of moisture from the material interior is

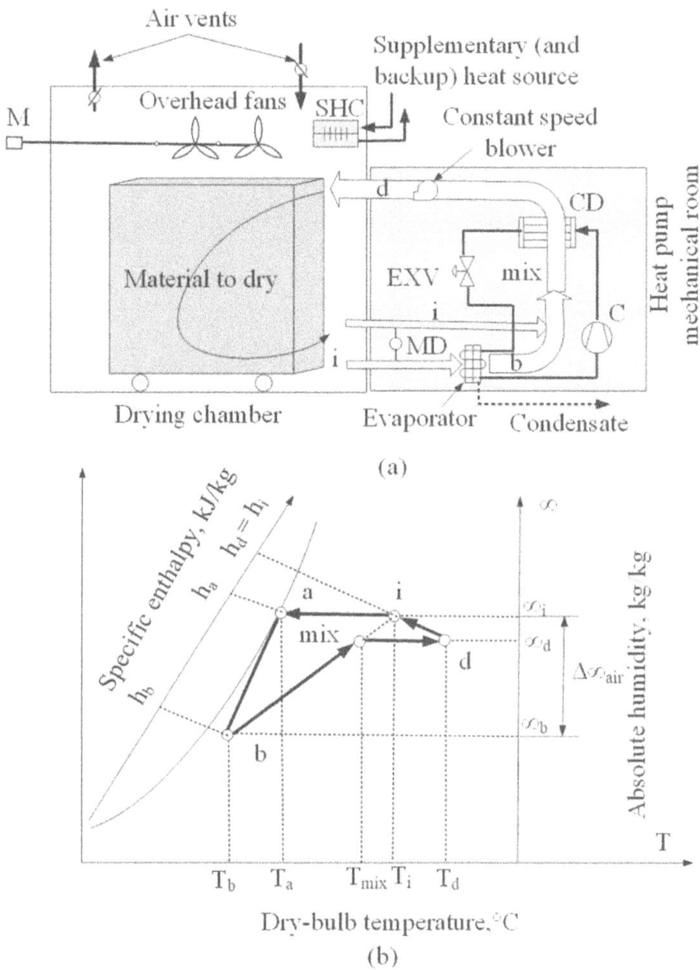

FIGURE 7.4 (a) Batch dryer coupled with *compact* mechanical vapor compression heat pump—intermittent drying with evaporator bypass and constant speed blower. (b) Thermodynamic processes of drying air represented in the Mollier ($h-T$) diagram. *C*, compressor; *CD*, condenser; *EXV*, expansion valve; *M*, external fan electrical motor; *MD*, motorized damper. *Note:* Schematic not to scale; not all components shown.

sufficient to preserve the liquid film on the material surface and, thus, to ensure an efficient operation when the compressor running is resumed (Minea 2006).

As can be seen in Figure 7.5, the heat pump compressor can be set to run according to the predetermined (preset) hourly percentage times, and the drying air mass flow rate (thus, the velocity) over the evaporator heat transfer surface is simultaneously increased from a minimum value (at the beginning) to a maximum value (toward the end of during cycle) (*Note:* Compressor hourly running time percentages up to the end of drying cycle not shown.) (Minea 2006).

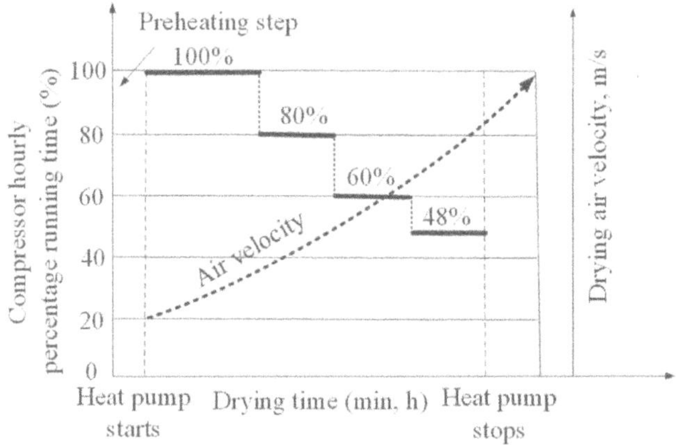

FIGURE 7.5 Compressor hourly (preset) percentage running times of heat pump-assisted drying systems with intermittent drying operation (see Figure 7.4). *Note:* Schematic not to scale; not all compressor hourly percentages running times shown.

The compressor running intermittency ratio (when the heat pump compressor is running) is defined as follows (Chua et al. 2002):

$$\alpha = \tau_{on} / \left(\tau_{on} + \tau_{off}\right) \tag{7.9}$$

where

α is the compressor running intermittency ratio (–);
τ_{on} is the compressor running duration (hours or minutes); and
τ_{off} is the compressor stopping duration (hours or minutes).

Table 7.1 shows an intermittent drying schedule performed with hardwood (yellow birch) in an experimental laboratory-scale heat pump-assisted batch dryer (Minea 2019). It can be seen that during the preheating step (when the compressor is not running), the heating agent (steam in this case) provided heat for 4.75 hours to increase the temperature of the dryer structure and equipment, as well as of the dried material stack. The heat pump started up when the drying air actual wet-bulb temperature attained its set value. After the preheating step, the heat pump ran intermittently for 210.25 hours, and the total drying cycle lasted 215 hours. In this example, the initial and the final average moisture content of the lumber stack varies from 75.9% (d.b.) to 7.6% (d.b.).

Based on the schedule shown in Table 7.1, Figure 7.6 represents the compressor preset hourly running times profile together with the set dry-bulb temperature and actual (measured) wet-bulb temperature (blue line).

Toward the end of drying cycle, the percentage of compressor running times may increase, as shown in Figure 7.6. Significant energy consumption savings and operating costs is possible through the employment of intermittent drying technique due to lower compressor electrical energy consumption during off (tempering) periods

TABLE 7.1
Drying Schedule of an Intermittent Heat Pump-Assisted Drying System

Drying Step	Time (hours)	Cumulative Time (hours)	Set Dry-Bulb Temperature (°C)	Set Wet-Bulb Temperature (°C)	Percentage of Compressor Hourly Running Time (%)
Preheating	4.75	0.00	37.7	32.2	0
2	71.75	4.75	37.7	32.2	100
3	29.50	76.50	40.5	38.9	85
4	24.00	106.00	43.3	34.4	75
5	12.33	130.00	46.1	34.4	65
6	30.83	142.33	54.4	34.4	65
7	16.58	173.17	57.2	35.0	75
8	3.50	189.75	60	35.0	85
9	21.75	193.25	60	35.0	100
Total time	–	215.00	–	–	–

when the moisture is allowed to migrate faster from inner cores of materials to their exposed surfaces.

The heat pump's compressor (and blower) ran only when the actual drying air wet-bulb temperature was higher than the preset values. When the drying air actual wet-bulb temperature (red line) became higher than its set value, the compressor hourly running time increased. On the contrary, when the drying air actual wet-bulb

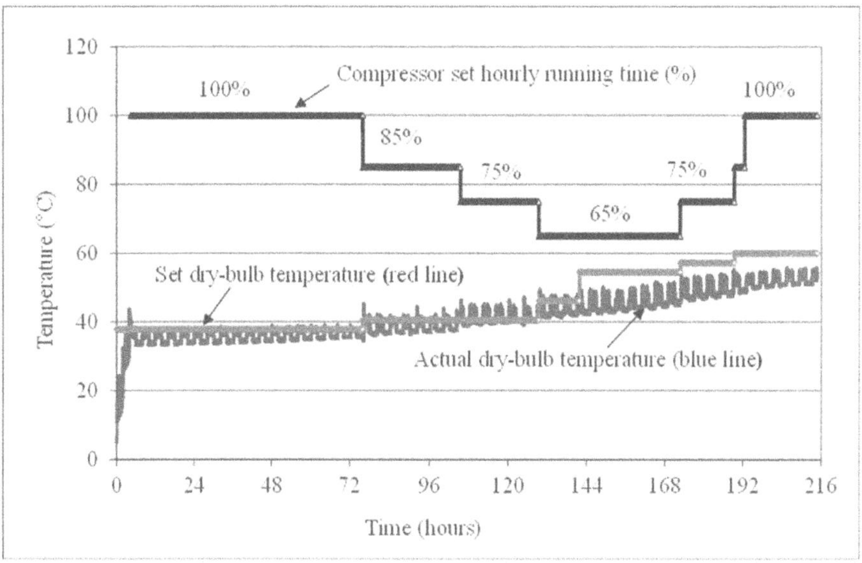

FIGURE 7.6 Example of compressor hourly (preset) percentages of running times (%), set dry-bulb temperature of drying air (red line) and actual (measured) dry-bulb temperature of drying air (blue line).

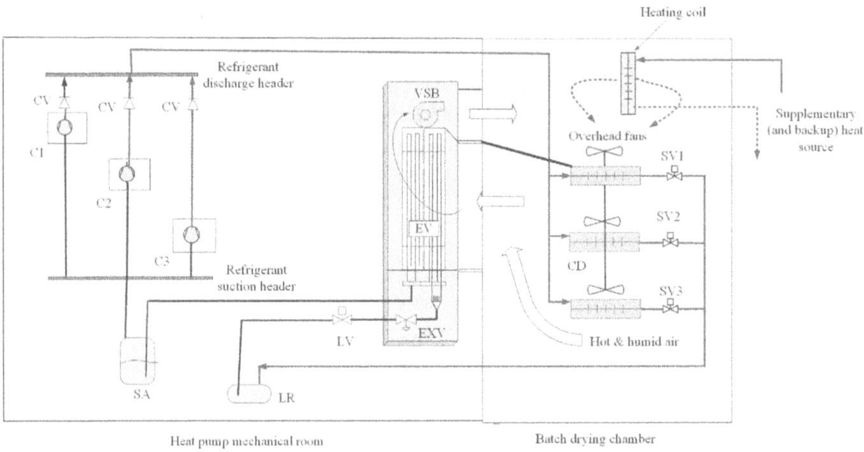

FIGURE 7.7 Schematic of a SPLIT heat pump-assisted drying batch drying system with multiple compressors and VSB. *C*, compressor; *CD*, remote condenser; *CV*, check valve; *SV*, solenoid valve; *VSB*, variable speed blower.

temperature approached its set value, the compressor hourly running time decreased (Minea 2019).

7.2.4 Multiple Heat Pump Compressors

Since the quantity of moisture extracted from dried materials gradually decreases from the beginning toward the end of drying cycles, the heat pumps' dehumidification capacities (i.e., evaporator thermal power) could be proportionally reduced by varying the number of multiple running compressors installed in parallel (Figure 7.7a and b).

The number of simultaneously running compressors should vary, for example, from maximum (e.g., 3–5) to at least one compressor. Such a strategy could be accompanied by varying the mass flow rate (and thus the velocity) of drying air over the system's common evaporator by using a VSB (Figure 7.7) (Minea 2015).

7.3 OPTIMUM CAPACITIES DESIGN (SIZE) OF MECHANICAL VAPOR COMPRESSION HEAT PUMPS COUPLED WITH HOT AIR CONVECTIVE BATCH DRYERS

Nominal thermal (dehumidification) capacities of mechanical vapor compression heat pumps used to dry porous and hygroscopic materials must be carefully sized.

Optimum design (size) capacities of mechanical vapor compression heat pumps coupled with hot air convective batch dryers require detailed preliminary information on the following (Mujumdar and Devahastin 2000): (i) physical properties of dried materials; (ii) material initial and final (required) moisture contents; (iii) material final quality requirements; and (iv) dryer optimum throughput.

Appropriate correlation between the material initial mass and its average moisture content (both determining the heat pump dehumidification capacity) and the moisture extraction (dewatering) rate from the product to dry should be always ensured. In other words, sufficient quantity of dried material has to be supplied to provide sufficient moisture extraction and thermal (latent and sensible) input to the heat pump evaporator.

Insufficient quantity and/or insufficient initial average moisture content of product to dry means (Minea 2019) (i) small quantity of moisture and associated latent heat extraction; (ii) too low compressor suction pressures and temperatures; (iii) risks of humidity freezing on the evaporator external fins; and (iv) too short drying cycles. In addition, such poor operating conditions may lead to several troubleshootings of mechanical vapor compression heat pumps.

To achieve high heat pump coefficients of performance $\left(COP_{heat\ pump}\right)$ (defined as the ratio of total energy delivered by the heat pump condenser and total electrical energy consumed by the heat pump compressor and blower) and system MSMER (defined as the ratio between the total mass of moisture extracted, expressed in kg_{water}, and the total energy consumed by the entire heat pump-assisted dryer, expressed in kWh), the dried material should give off enough moisture to be able to provide sufficiently high absolute humidity to the drying air at the heat pump evaporator inlet. This ensures that enough condensing enthalpy (latent heat) will be recovered. Both over- and undersized heat pump capacities may drastically penalize the heat pump coefficients of performance ($COP_{heat\ pump}$) and the system MSMER. Such a challenge could be overcome only if the heat pump dehumidification capacity (evaporator thermal power) is sized as an average value between the estimated maximum and minimum material dewatering rates. This means that the quantity of moisture removed from the material stacks (e.g., fruits, vegetables, or woods) should match the heat pump dehumidification capacity (i.e., evaporator thermal power, which naturally varies according to thermal energy recovered from the drying air) during the entire drying cycle. Such an approach helps to keep the heat pump compressor in practically continuous running mode.

To achieve as high as possible $COP_{heat\ pump}$ and $MSMERs$, the heat pump-assisted dryers should be designed to provide sufficient absolute humidity of the drying air at the heat pump evaporator inlet. This ensures high-enough condensing enthalpy (latent heat) to be recovered by the heat pump evaporator in order to provide high drying performances and avoid intermittent operation.

The heat pump dehumidification capacity (evaporator thermal power) must be as low as possible according to the material characteristics and its initial quantity of moisture (water). Such an approach allows to not oversize the heat pump compressor and, thus, to avoid frequent and short on/off running cycles. In other words, an optimum ratio between the maximum quantity of moisture that can be removed from the material stack(s) and the heat pump maximum and minimum dehumidification capacity has to be provided, while keeping the heat pump in quasi-continuous running mode and without rejecting excess heat outdoors.

Optimum design (sizing) point of mechanical vapor compression heat pumps coupled with convective hot air batch dryers depends, among many other parameters, on (estimated) maximum, minimum, and average values of heat pump evaporator thermal power (dehumidification capacity) that vary from the beginning toward the end of drying cycles. Therefore, the designers of heat pump-assisted drying systems

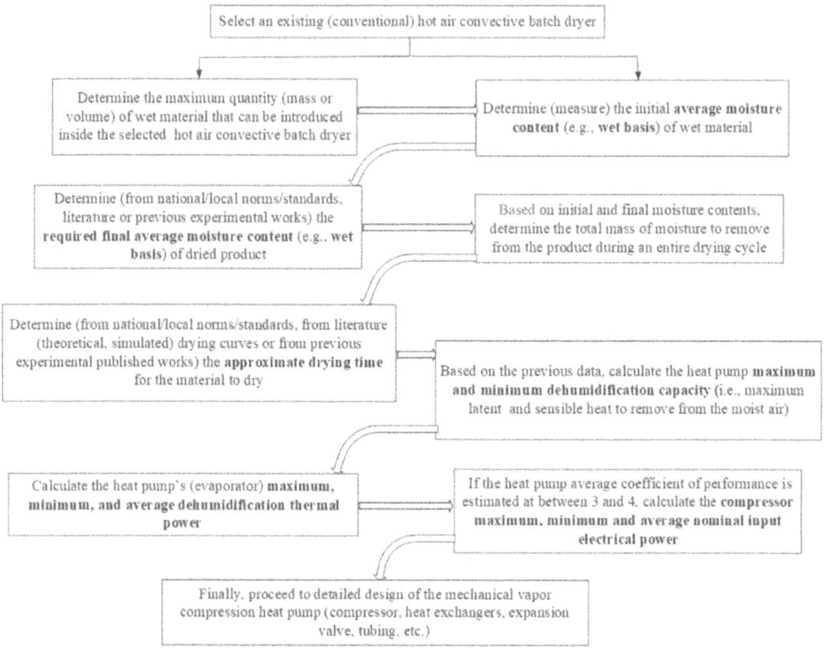

FIGURE 7.8 The main steps of design process of heat pump-assisted batch dryers.

must first evaluate (estimate) these thermal parameters to efficiently dry products such as fruits, vegetables, and woods.

The following are the main steps of optimum design (size) process, as shown in Figure 7.8 (Minea 2019):

1. Select an existing (conventional) hot air convective batch dryer designed for drying a given product (e.g., fruits, vegetables, or woods).
2. Determine the initial maximum quantity (mass or volume) of the wet material (product) that can be introduced (stored) inside the selected conventional (existing) hot air convective batch dryer ($m_{wet\ material}^{initial,max}$) ($kg_{wet\ material}$ or $m^3_{wet\ material}$).
3. Measure (determine) the initial average moisture content (e.g., wet basis) of the initial quantity of the wet material (product) ($\overline{MC}_{in}$) ($kg_{water}/kg_{wet\ material} = kg_{water}/(kg_{water} + kg_{dry\ material})$.
4. Determine (from national/local norms/standards, literature, or previous experimental works) the required final average moisture content (e.g., wet basis) of the dried product ($\overline{MC}_{fin}$) ($kg_{water}/kg_{wet\ material} = kg_{water}/(kg_{water} + kg_{dry\ material})$.
5. Based on $\overline{MC}_{in}$ and $\overline{MC}_{fin}$, determine the total mass (weight) of moisture (water) to remove from the product during an entire drying cycle ($m_{moisture}^{removed}$) (kg_{water}).
6. Determine (from national/local norms/standards, from literature [theoretical, simulated] drying curves or from previous experimental published works) the required (approximate) drying time (t_{drying}).

7. Based on the previous data, calculate (estimate) the heat pump evaporator maximum and minimum dehumidification capacity (i.e., maximum latent and sensible heat to remove) ($Q_{heat\ to\ remove}^{evaporator}$) (kJ).
8. Calculate the evaporator maximum ($\dot{Q}_{dehum}^{max}$) and minimum ($\dot{Q}_{dehum}^{min}$) dehumidification thermal power (kW).
9. Assuming a heat pump average coefficient of performance $\left(\overline{COP}_{heat\ pump}\right)$ between 3 and 4, calculate the compressor maximum and minimum input electrical power ($\dot{W}_{input\ pwer}^{compressor}$) (kW).
10. Finally, proceed to detailed design of the mechanical vapor compression heat pump (compressors, heat exchangers, expansion valve, tubing).

7.3.1 As Function of Maximum and Minimum Estimated Absolute Humidity of Drying Air

As can be seen in Figure 7.9a, the optimum location (not shown to scale) of the nominal design point of mechanical vapor compression heat pumps coupled with hot air convective batch dryers could be determined as function of an (estimated) average value of absolute humidity (or wet-bulb temperature) of the drying air over the entire drying cycle. It can be seen that the optimum heat pump design point should be located somewhere between the estimated maximum (at the beginning of the drying cycle) and the estimated minimum (at the end of the drying cycle) values of drying air absolute humidity (or wet-bulb temperatures).

This (average) optimum design capacity of mechanical vapor compression heat pump could be determined (estimated) according to the material initial and (required) final average moisture contents, thus as an average of estimated maximum and minimum absolute humidity (or wet-bulb temperatures) of the drying air inside the heat pump-assisted batch dryers (Minea 2015).

Figure 7.9b shows, as an example, set and actual (measured) wet-bulb temperatures of drying air in the case of dried softwood lumber boards (Minea 2019).

Such a design method is possible because the mechanical vapor compression heat pumps are able to automatically (naturally) adjust their dehumidification rates (evaporator thermal powers and corresponding compressor electrical power inputs) between two extreme operation values:

$$\dot{Q}_{dehum}^{min} \lesssim \dot{Q}_{dehum}^{nominal} \lesssim \dot{Q}_{dehum}^{max} \tag{7.10}$$

$$\dot{W}_{compr}^{min} \lesssim \dot{W}_{compr}^{nominal} \lesssim \dot{W}_{compr}^{max} \tag{7.11}$$

where

$\dot{Q}_{dehum}^{min}$ is the heat pump minimum dehumidification capacity (i.e., evaporator thermal power) corresponding to the minimum absolute humidity (or wet-bulb temperature) of drying air (kW);

$\dot{Q}_{dehum}^{nominal}$ is the heat pump nominal dehumidification capacity (i.e., evaporator nominal thermal power) (kW);

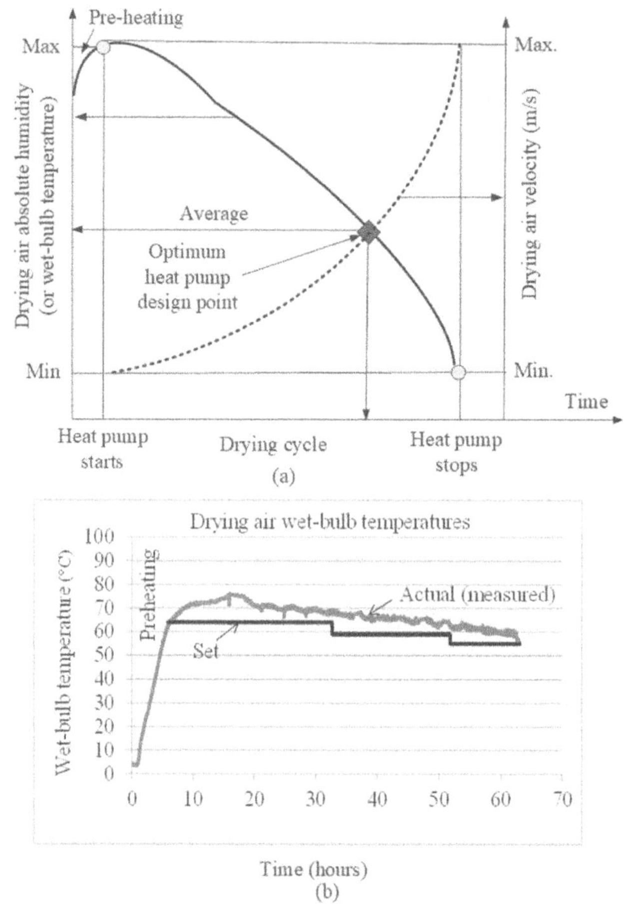

FIGURE 7.9 (a) Optimum location of heat pump capacity design point as function of estimated (average) absolute humidity (or wet-bulb temperature) of drying air. *Note:* Graph not to scale. (b) Example of set and actual (measured) wet-bulb temperatures of drying air in the case of dried softwood lumber boards. (From Minea 2019.)

$\dot{Q}_{dehum}^{max}$ is the heat pump maximum dehumidification capacity (i.e., evaporator maximum thermal power) corresponding to the maximum absolute humidity (or wet-bulb temperature) of drying air (kW);

$\dot{W}_{compr}^{min}$ is the compressor minimum electrical input pawer corresponding to the minimum absolute humidity (or wet-bulb temperature) of drying air (kW);

$\dot{W}_{compr}^{nominal}$ is the compressor nominal electrical input power (kW); and

$\dot{W}_{compr}^{max}$ is the compressor maximum electrical input pawer corresponding to the maximum absolute humidity (or wet-bulb temperature) of drying air (kW).

Equations 7.11 and 7.12 show that when the absolute humidity of the drying air is at its maximum value (at the beginning of drying cycle), the heat pump dehumidification rate (and the corresponding compressor electrical power input) will attain its maximum ($\dot{Q}_{dehum}^{max}$). When the absolute humidity of the drying air is at its minimum value (at the end of drying cycle), the heat pump dehumidification rate (and the corresponding compressor electrical power input) will attain its minimum value ($\dot{Q}_{dehum}^{min}$).

In order to match the heat pump's actual dehumidification capacity (i.e., evaporator thermal power) with the material dewatering rate), the velocity of drying air must be increased from a minimum value (at the beginning of drying cycles) to a maximum value (toward the end of drying cycles) in both *compact* and *split* heat pump-assisted drying systems as shown in Figure 7.9a.

7.3.2 As Function of Maximum and Minimum Estimated Mass of Moisture to Remove and Drying Air Variable Velocity

Because the external heat exchange surfaces of heat pump evaporators are constant, it is necessary to vary the drying air flow rates (and, thus, velocities) to keep the evaporator thermal power (dehumidification capacity) matched as much as possible with materials' dewatering rates.

When the maximum and minimum moisture contents of dried materials can be estimated, the heat pump optimum (nominal) design point could be determined as a function of the average mass of moisture to remove (Figure 7.10a). In this case,

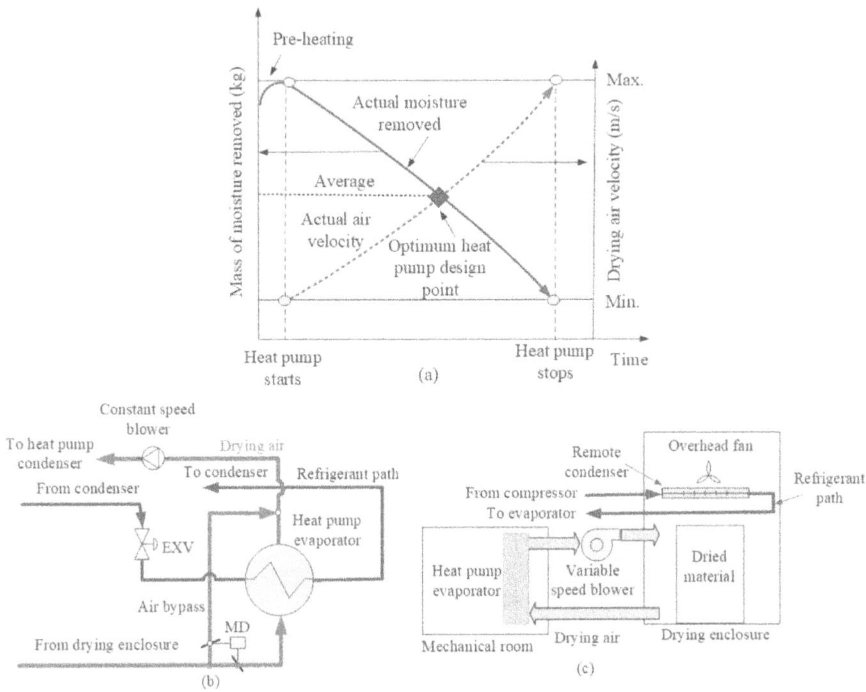

FIGURE 7.10 (a) Optimum location of heat pump capacity design point as function of maximum and minimum estimated mass of moisture to remove and variable drying air velocity. (b) Case of compact-type heat pumps. (c) Case of split-type heat pumps. *Note:* Schematics not to scale. EXV, expansion valve; MD, motorized damper.

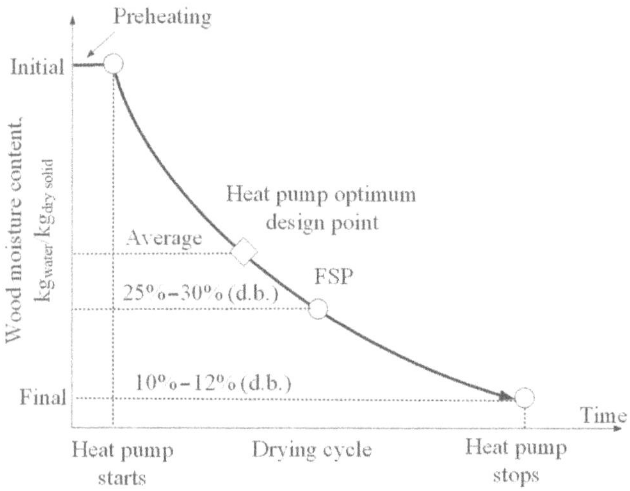

FIGURE 7.11 Relative position of heat pump nominal design (size) point as function of actual wood moisture content. *Note:* Graphic not to scale. *FSP*, fiber saturation point.

in order to match (correlate) the heat pumps' actual dehumidification capacity (i.e., evaporator thermal power) to the material dewatering rate, the velocity of drying air must be simultaneously increased from a minimum value (at the beginning of drying cycles) to a maximum value (toward the end of drying cycles) in both compact (Figure 7.10b) and split (Figure 7.10c) heat pump-assisted drying systems.

Figure 7.11 suggests that in the case of wood heat pump-assisted dryers, optimum location of the heat pump nominal design point should be situated at (or just before) the material fiber saturation point (*FSP*), defined as the moisture content at which the cells' cavities are completely devoid of water, but the cells' walls are still completely saturated. For softwood species, fiber saturation point is of between 25% and 30% (d.b.).

7.3.3 As Function of Maximum and Minimum Estimated Evaporator Thermal Power and Drying Air Variable Velocity

As shown in Section 7.3.2, because the heat pump's evaporator external heat exchange surface is constant, it is necessary to vary the drying air velocity (or flow rate) to keep dehumidification rates (evaporator thermal power) matched (correlated) as much as possible with the material dewatering rate.

When the maximum and minimum evaporator thermal powers can be estimated, the heat pump optimum (nominal) design point could be determined (estimated) as an average value (Figure 7.12a). In this case, in order to match the heat pump's actual dehumidification capacity (i.e., evaporator thermal power) with the material dewatering rate, the velocity of drying air must be increased from a minimum value (at the beginning of drying cycles) to a maximum value (toward the end of drying cycles) in both *compact* (Figure 7.12b) and *split* (Figure 7.12c) heat pump-assisted drying systems.

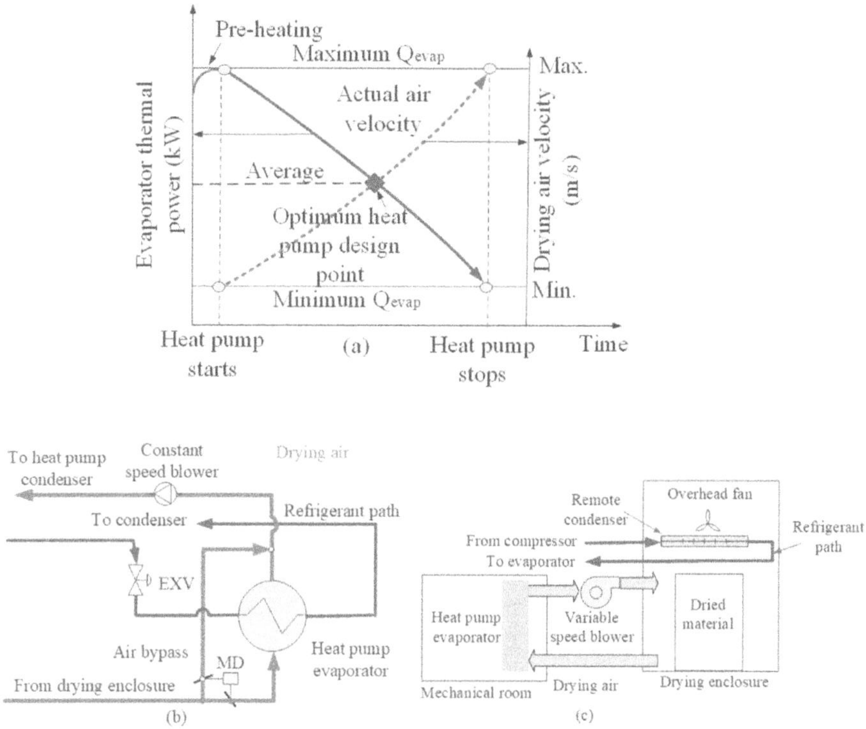

FIGURE 7.12 (a) Optimum location of heat pump capacity design point as function of maximum and minimum estimated evaporator thermal power and variable drying air velocity. (b) Case of compact-type heat pumps. (c) Case of split-type heat pumps. *Note:* Schematics not to scale. EXV, expansion valve; MD, motorized damper.

7.3.4 As Function of Maximum and Minimum Estimated Compressor Electrical Power Inputs and Drying Air Variable Velocity

If maximum and minimum evaporator thermal power (i.e., heat pump dehumidification capacity) can be estimated, the maximum and minimum compressor electrical power input could also be estimated by assuming, for example, heat pump average coefficients of performance ($COP_{heat\ pump}$) between 3 and 4. In this case, the optimum location of heat pump capacity design (size) point could be located as an average point between the maximum and minimum electrical input power (Figure 7.13a) for both *compact* (Figure 7.13b) and *split* (Figure 7.13c) heat pump types.

It can be seen that to remove the moisture evaporated from the dried product (which is maximum at the beginning of drying cycle and minimum at the end of drying cycle), the drying air velocity over the evaporator external heat transfer surface should (must) be proportionally increased from minimum to maximum appropriate values (Figure 7.13a).

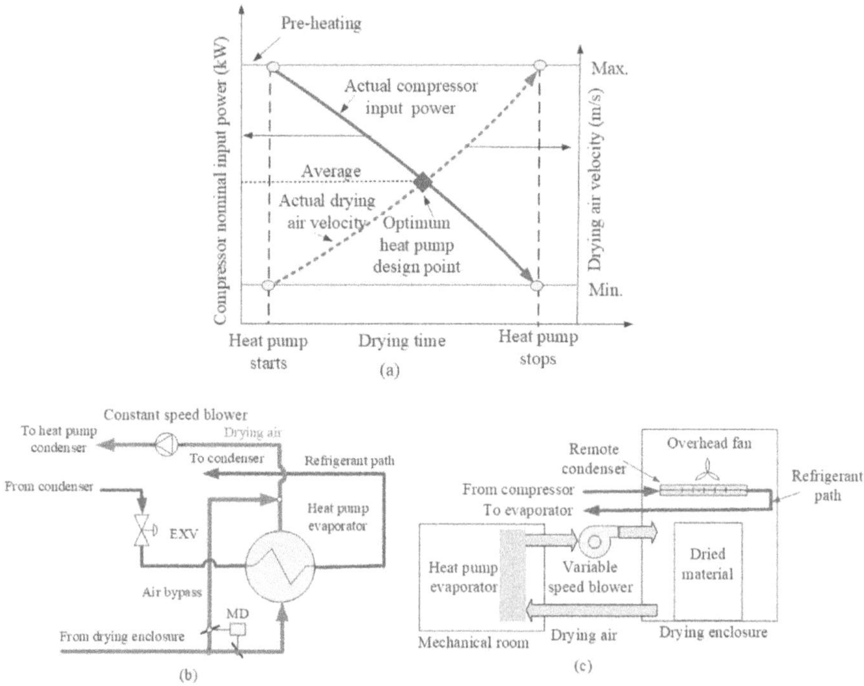

FIGURE 7.13 (a) Optimum location of capacity heat pump design point as function of maximum and minimum estimated compressor electrical power input and variable drying air velocity. (b) Case of compact-type heat pumps. (c) Case of split-type heat pumps. *Note:* Schematics not to scale. EXV, expansion valve; MD, motorized damper.

7.3.5 As Function of Estimated Compressor Speed and Drying Air Velocity

When variable speed compressors are used for mechanical vapor compression heat pumps coupled with hot air convective batch dryers, the location of optimum heat pump capacity design (size) point could be found as an average value of the compressor speed (which should be maximum at the beginning and minimum at the end of drying cycle) (Figure 7.14). This operating rule is required since during drying cycles, the heat pump dehumidification capacity (evaporator thermal power) has to be continuously varied and adapted to the actual quantity of moisture removed that should match the material actual thermal dewatering capacity.

It can be seen that to be able to remove the moisture evaporated from the dried product (which is maximum at the beginning of drying cycle and minimum at the end of drying cycle), the drying air velocity over the evaporator external heat transfer surface should (must) be proportionally increased from minimum to maximum appropriate values.

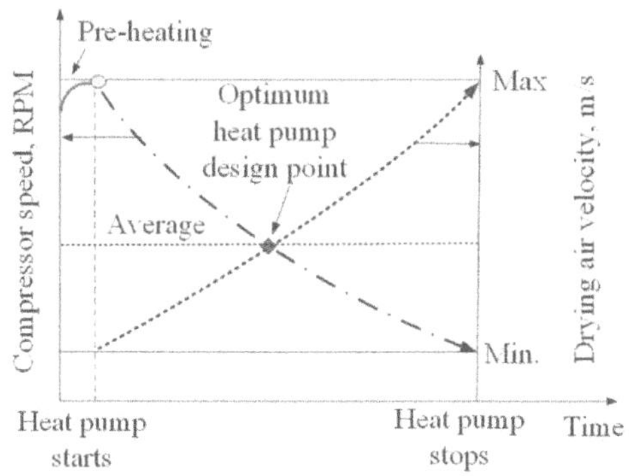

FIGURE 7.14 Optimum heat pump capacity design point as function of maximum and minimum compressor speed and variable drying air velocity over evaporator external heat transfer surface. *Note:* Graph not to scale.

7.3.6 As Function of Estimated Maximum and Minimum Number of Running Compressors and Drying Air Velocity

Instead of using a single variable speed compressor, if multiple parallel constant speed compressors are employed, then the average number of running compressors (and thus the optimal average design points of mechanical vapor compression heat pumps coupled with hot air convective batch dryers) could be determined as an average between the maximum and minimum number of simultaneously running compressors (Figure 7.15).

It can be seen that to remove the moisture evaporated from the dried product (which is maximum at the beginning of drying cycle and minimum at the end of

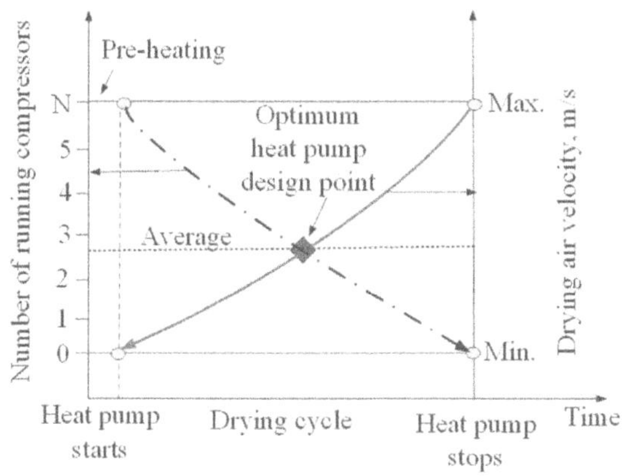

FIGURE 7.15 Optimum heat pump capacity design point as function of estimated mass of moisture to remove, number of running compressors, and variable drying air velocity over evaporator external heat transfer surface. *Note:* Schematic not to scale.

drying cycle), the drying air velocity over the evaporator external heat transfer surface should be proportionally increased from minimum to maximum appropriate values.

7.3.7 Solved Exercises

7.3.7.1 Drying Mango Slices

a. **Given Data**

- Product to dry: mango slices
- Initial (maximum) mass of the wet material: $m_{wet\ material}^{initial,max} = 1,000\ \text{kg}$
- Average initial moisture content (w.b.) of wet product: $\overline{MC}_{in} = 87.2\%$ $\text{kg}_{water}/kg_{wet\ material}$)
- Average final moisture content (w.b.) of dried product: $\overline{MC}_{fin} =$ $8.7\%\ \text{kg}_{water}/\text{kg}_{wet\ material}$
- Approximate drying time: $t_{drying} = 12\ \text{hours}$
- Moisture (water) condensing (latent heat) enthalpy: $h_{water}^{condensing} =$ $2{,}260\ \text{kJ/kg}_{water}$
- Heat pump average coefficient of performance: $\overline{COP}_{heat\ pump} \simeq 3.5$

b. **Determine**

- Initial mass of moisture (water) contained inside the initial mass of the wet product
- Final mass of moisture (water) contained inside final mass of dried product
- Total mass of moisture (water) removed from the dried product during the entire drying cycle
- Total latent heat removed by heat pump's evaporator from the initial mass of wet product during the entire drying cycle
- Average (nominal) heat pump dehumidification capacity (evaporator thermal power)
- Average (nominal) electrical power input of heat pump compressor

c. **Solution**

- Initial mass of moisture (water) contained inside the initial mass of wet product:

$$m_{moisture}^{initial} = \frac{\overline{MC}_{in}}{100} \cdot m_{wet\ material}^{initial,max} = \frac{87.2}{100} \times 1{,}000 = 872\ \text{kg}_{moisture}^{initial}$$

- Final mass of moisture (water) contained inside final mass of dried product:

$$m_{moisture}^{final} = \frac{\overline{MC}_{fin}}{100} \cdot m_{wet\ material}^{final} = \frac{8.7}{100} \times 1{,}000 = 87\ \text{kg}_{moisture}^{final}$$

- Total mass of moisture (water) *removed* from the initial mass of wet product during the entire drying cycle:

$$m_{moisture\ removed}^{total} = m_{moisture}^{initial} - m_{moisture}^{final} = 872\ \text{kg}_{moisture}^{initial} - 87\ \text{kg}_{moisture}^{final}$$
$$= 775\ \text{kg}_{moisture}$$

- Total latent heat removed by heat pump's evaporator from the initial mass of wet product during the entire drying cycle:

$$Q_{latent}^{total} = m_{moisture\,removed}^{total} \cdot h_{water}^{condensing} = 785\,\text{kg}_{water} \cdot 2,260 \frac{\text{kJ}}{\text{kg}_{water}}$$
$$= 1.774 \times 10^6\ \text{kJ}$$

- Average heat pump dehumidification capacity (evaporator thermal power):

$$\dot{Q}_{evap}^{dehum} = \frac{Q_{latent}^{total}}{t_{drying}} = 1.774 \times 10^6\ \text{kJ}/\left(12 \times 3,600\,\text{s}\right) = 41.06\,\text{kW}$$

- Design (nominal) heat pump dehumidification capacity (nominal evaporator thermal power):

$$\dot{Q}_{evap,nominal}^{dehum} = \dot{Q}_{evap}^{dehum}/2.5 = 41.06\,\text{kW}/2.5 = 16.42\,\text{kW}$$

- Average (nominal) electrical power input of heat pump compressor:

$$\dot{W}_{evap}^{dehum} = \dot{Q}_{evap,nominal}^{dehum}/\left(\overline{COP}_{heat\ pump} - 1\right) = 16.42/(3.5-1) = 6.56\,\text{kW}$$

7.3.7.2 Drying Lumber Boards

a. **Given Data**
- Consider $m_{softwood} = 6,700\,\text{kg}$ of green (wet) softwood dried in a heat pump-assisted batch dryer
- Initial average moisture content: $\overline{MC}_{in} = 80\%(\text{w.b.})$
- Final average moisture content: $\overline{MC}_{fin} = 18\%(\text{w.b.})$
- Estimated drying cycle time: $t_{drying} = 60\,\text{hours}$
- Average heat pump coefficient of performance: $\overline{COP}_{heat\ pump} = 3.5$

b. **Determine**
- Heat pump nominal dehumidification capacity (evaporator nominal thermal power)
- Compressor nominal electrical power input

c. **Solution**
- Softwood initial moisture mass:

$$m_{moisture}^{initial} = \left(\overline{MC}_{in}/100\right) \cdot m_{softwood} = \left(\frac{80}{100}\right) \times 6,700 = 5,360\,\text{kg}_{moisture}^{initial}$$

- Softwood final moisture mass:

$$m_{moisture}^{final} = \left(\overline{MC}_{fin}/100\right) \cdot m_{softwood} = \left(\frac{18}{100}\right) \times 6,700 = 1,206\,\text{kg}_{moisture}^{final}$$

- Total mass of moisture removed:

$$m_{moisture}^{removed} = m_{moisture}^{initial} - m_{moisture}^{final} = 5,360\,\text{kg}_{moisture}^{initial} - 1,206\,\text{kg}_{moisture}^{final}$$
$$= 4,154\,\text{kg}_{moisture}^{removed}$$

- Latent heat removed by heat pump evaporator:

$$Q_{latent\ heat}^{removed} = 0.9 \times \left(m_{moisture}^{removed} \cdot h_{water}^{condensing} \right)$$

$$= 0.9 \times \left(4{,}154\,\text{kg}_{moisture}^{removed} \times 2{,}260 \frac{\text{kJ}}{\text{kg}_{water}} \right) = 8.46 \times 10^6\ \text{kJ}$$

- Sensible heat removed by heat pump evaporator represents approximately 30% of latent heat removed:

$$Q_{sensible\ heat}^{removed} = 0.3 \cdot Q_{latent\ heat}^{removed} = 0.3 \times \left(8.46 \times 10^6\ \text{kJ} \right) = 2.54 \times 10^6\ \text{kJ}$$

- Total heat (sensible plus latent) removed by heat pump evaporator:

$$Q_{total\ heat}^{removed} = Q_{latent\ heat}^{removed} + Q_{sensible\ heat}^{removed} = 8.46 \times 10^6\ \text{kJ} + 2.54 \times 10^6\ \text{kJ}$$

$$= 11 \times 10^6\ \text{kJ}$$

- Average thermal power removed by heat pump evaporator:

$$\overline{\dot{Q}}_{average}^{removed} = \frac{Q_{total\ heat}^{removed}}{t_{drying}} = 11 \times 10^6\ \text{kJ} / \left(60 \times 3{,}600\,\text{s} \right) = 50.9\,\text{kW}$$

- Optimum design heat pump dehumidification capacity (evaporator nominal thermal power):

$$\overline{Q}_{evaporator}^{nominal} \simeq \overline{\dot{Q}}_{average}^{removed} / 2.5 = 50.9\ \text{kW} / 2.5 = 20.37\,\text{kW}$$

- Optimum design (nominal) compressor electrical power input:

$$\dot{W}_{compressor}^{nominal} = \overline{Q}_{evaporator}^{nominal} \left(\overline{COP}_{heat\ pump} - 1 \right) = 20.37\,\text{kW} \cdot (3.5 - 1) = 50.9\,\text{kW}$$

7.4 MOISTURE CONDENSERS (REFRIGERANT EVAPORATORS)

Heat exchangers operating with humid air as heat sources and refrigerants as heat sink sources are called moisture condensers. In such external finned-tube heat exchangers, some of the moisture present in the moist air condense (or even freeze) over the outside fins and tube surfaces, while inside the tubes, partial or total vaporization of refrigerants occurs.

Moisture condensers are critical components of air-to-air mechanical vapor compression heat pumps coupled with hot air convective batch dryers. They are designed by assuming steady-state conditions with nonhomogeneous, two-phase refrigerant flow inside tubes and crossflow drying air outside (McQuiston et al. 2005).

The overall heat transfer coefficients of finned-tube moisture condensers depend on sensible and latent heat transfers on both air-side (from hot moist air to the external surfaces) and refrigerant-side (from the tube internal wall to the vaporizing refrigerant) (Figures 7.16a–c).

In heat pump-assisted dryers, the moisture condensers operate partially under external dry-surface conditions during sensible heat recovery from the drying air (process $1-a$ in Figure 7.16c) and refrigerant superheating (process $B-C$ in Figure 7.16b)

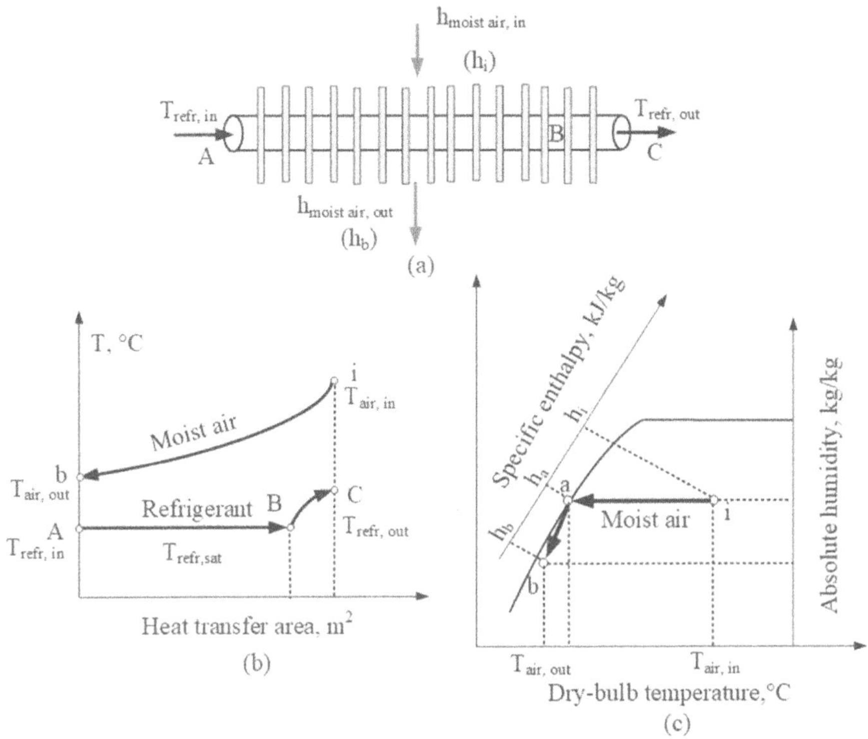

FIGURE 7.16 (a) Schematic of a finned-tube moisture condenser tube (heat pump evaporator). *Note:* Internal and external fouling layers are not shown; schematic not to scale. (b) Temperature variations of moist air and refrigerant represented in a $T - A$ diagram. (c) Cooling and dehumidification processes represented in the Mollier ($h - T$) diagram. *sat*, saturation.

and, partially, under external wet surface conditions during the moisture condensation (process $a - b$ in Figure 7.16c).

On the air-side of moisture condensers, both cooling and dehumidification processes consisting in heat and mass (water) transfers occur simultaneously. The state of the moist air (state 1 in Figure 7.16b and c) moves toward the saturation curve (100% relative humidity) of which average temperature is that of the wetted (tube and fins) surfaces. Typically, the air-dry bulb temperature and the absolute humidity decrease as the air flows over the heat exchanger. Therefore, sensible and latent heat transfers occur simultaneously (McQuiston et al. 2005).

During normal heat exchanger operation of moisture condensers, surfaces are often subject to fouling by fluid impurities, rust formation, or other reactions between the fluid and the wall material. The subsequent deposition of a film or scale on the surface can greatly increase the resistance to heat transfer between the fluids. This effect can be treated by introducing additional thermal resistances (termed as fouling factors) that depend on operating temperatures and fluid velocities. In practice, the total thermal resistances comprise air-side resistance, air-side fouling resistance, tube wall, fin-to-tube contact, refrigerant-side fouling, and refrigerant-side resistances. Fouling factors are variable during heat exchanger operation (increasing from zero for a clean surface), as deposits accumulate on both external and internal surfaces (Incropera and DeWitt 2002).

The overall (mean) thermal resistance, including the fouling factors of extended surfaces (tubes and fins), could be expressed as follows (Incropera and DeWitt 2002; McQuiston and Parker 1988):

$$R_{tot} = \frac{1}{\bar{U}} = \frac{1}{\eta_{air} \cdot h_{air} \cdot A_{air}} + \frac{1}{\eta_{air} \cdot h_{a,f} \cdot A_{air}} + \frac{1}{R_c \cdot A_c} + \frac{1}{2\pi \cdot k_{wall} \cdot L} \ln\frac{d_{ext}}{d_{int}} + \frac{1}{\eta_{refr} \cdot h_{r,f} \cdot A_{refr}} + \frac{1}{\eta_{refr} \cdot h_{refr} \cdot A_{refr}} \tag{7.12}$$

where

R_{tot} is the overall (mean, total) thermal resistance ($m^2 \cdot K/W$);
$\bar{U}$ is the overall (mean, total) heat transfer coefficient ($W/m^2 \cdot K$);
η_{air} is the air-side overall surface efficiency of enhanced finned tubes (–); for industrial in-use coils, η_{air} generally range between 0.3 and 0.7;
h_{air} is the air-side total heat transfer coefficient accounting for composed sensible and latent heat transfer coefficients (W/m^2K);
A_{air} is the air-side heat transfer area (m^2);
$h_{a,f}$ is the air-side fouling factor (heat conductance) (W/m^2K); for industrial moisture condensers in operation for more than five years, the air-side fouling factor may be of $2{,}840\ W/m^2K$;
R_c is the thermal contact conductance between the tube and fin surfaces in contact where a temperature drop exists; it is defined as the ratio between this temperature drop and the average heat transfer density across the interface (W/m^2K); the contact resistance between moisture condenser's tubes and fins (that can range between $10{,}000\ W/m^2K$ and $16{,}000\ W/m^2K$, with mean values of around $13{,}000\ W/m^2K$) appears because the heat is exchanged across an interface where two surfaces are in imperfect contact;
A_c is the contact area between fins and tubes (m^2);
k_{wall} is tube wall thermal conductivity (W/mK);
L is the total length of circular tubes (m);
d_{ext} is the external diameter of circular tubes (m);
d_{int} is the internal diameter of circular tubes (m);
η_{refr} is the refrigerant-side overall surface efficiency (–);
$h_{r,f}$ is the refrigerant-side fouling factor (conductance) (W/m^2K);
h_{refr} is the refrigerant-side heat transfer coefficient (W/m^2K); and
A_{refr} is the refrigerant-side tube surface area (m^2).

In Equation 7.14, the wall thermal conduction resistance term $\left(\frac{1}{2\pi \cdot k_{wall} \cdot L} \ln\frac{d_{ext}}{d_{int}}\right)$ can be neglected, since a thin wall of large thermal conductivity is generally used. Also, the air-side heat transfer coefficient is much smaller compared with that of refrigerant liquid–vapor mixture experiencing boiling, this justifies the usage of fins to enhance air-side convection heat transfer rates.

The calculation of overall heat transfer coefficients depends on whether they are based on the air-side or refrigerant-side surface area, since if $A_{refr} \neq A_{air}$, then $U_{refr} \neq U_{air}$.

Based on the air-side heat transfer area (A_{air}), Equation 7.12 gives

$$\bar{U}_{air} = 1 / \left[\frac{1}{\eta_{air} \cdot h_{air}} + \frac{1}{\eta_{air} \cdot h_{a,f}} + \frac{A_{air}}{R_c A_c} + \frac{A_{air} \cdot \ln(d_{ext} / d_{int})}{2\pi \cdot L \cdot k_{wall}} + \frac{1}{\eta_{refr} \cdot h_{r,f} (A_{refr} / A_{air})} + \frac{1}{\eta_{refr} \cdot h_{refr} (A_{refr} / A_{air})} \right] \tag{7.13}$$

where

$\bar{U}_{air}$ is the overall (mean, total) heat transfer coefficient based on the air-side heat transfer area (W/m^2).

Thermal power of moisture condensers can be determined as follows (see Figure 7.16a):

$$\dot{Q}_{moist,cond.} = \eta_{air} \cdot \dot{m}_{air} \cdot (h_{air,in} - h_{air,out}) = \dot{m}_{refr} \cdot (h_{refr,out} - h_{refr,in}) \tag{7.14}$$

where

$\dot{Q}_{moist,cond.}$ is the moisture condenser heat transfer rate (kW);
η_{air} is the air-side overall surface efficiency of enhanced finned-tubes (–); for industrial in-use coils, η_{air} generally range between 0.3 and 0.7 (–);
$\dot{m}_{air}$ is the humid air mass flow rate (kg/s);
$h_{air,in}$ is the specific enthalpy of moist air at moisture condenser inlet (kJ/kg);
$h_{air,out}$ is the specific enthalpy of moist air at moisture condenser outlet (kJ/kg);
$\dot{m}_{refr}$ is the refrigerant mass flow rate (kg/s);
$h_{refr,out}$ is the specific enthalpy of refrigerant at moisture condenser outlet (kJ/kg); and
$h_{refr,in}$ is the specific enthalpy of refrigerant at moisture condenser inlet (kJ/kg).

Thermal power of moisture condensers can be also expressed as follows:

$$\dot{Q}_{moist,cond.} \approx \bar{U}_{air} \cdot A_{air} \cdot LMTD \tag{7.15}$$

where

$\dot{Q}_{moist,cond.}$ is the moisture condenser heat transfer rate (kW);
$\bar{U}_{air}$ is the overall (mean) heat transfer coefficient (W/m$^2 \cdot$K) (see Equation 7.13);
A_{air} is the total air-side heat transfer area of moisture condenser (m^2);
$LMTD = (T_{air,in} - T_{air,out}) / \left[\ln(T_{air,in} - T_{refr}) / (T_{air,out} - T_{refr}) \right]$ is the logarithmic mean temperature difference (°C) (see Figure 7.16b);
$T_{air,in}$ is the moist air inlet temperature (°C);

$T_{air,out}$ is the moist air outlet temperature (°C); and
T_{refr} is the refrigerant evaporating (constant) temperature (°C).

By accounting both sides fouling and using Equation 7.12, the moisture condenser (heat pump evaporator) total heat transfer rate (kW) can be calculated as follows:

$$\dot{Q}_{moist,cond.} = \frac{LMTD \cdot A_{air}}{\frac{1}{h_{air} \cdot A_{air} \cdot \eta_{air}} + \frac{1}{\eta_{air} \cdot h_{a,f} \cdot A_{air}} + \frac{1}{R_c \cdot A_c} + \frac{1}{2\pi \cdot k_{all} \cdot L} ln \frac{d_{ex}}{d_{int}} + \frac{1}{\eta_{refr} \cdot h_{r,f} \cdot A_{refr}} + \frac{1}{\eta_{refr} \cdot h_{refr} \cdot A_{refr}}} \quad (7.16)$$

where
$\dot{Q}_{moist,cond.}$ is the total moisture condenser heat transfer rate (kW);
A_{air} is the air-side heat transfer area (m^2);
$LMTD$ is the logarithmic mean temperature difference (°C);
$h_{air} = h_{sens} + h_{lat}$ is the air-side total heat transfer coefficient accounting for composed sensible and latent heat transfer coefficients ($W/m^2 \cdot K$);
h_{sens} is the air-side sensible heat transfer coefficient ($W/m^2 \cdot K$); and
h_{lat} is the air-side latent heat transfer coefficient ($W/m^2 \cdot K$).

Based on Equation 7.16, the required heat transfer area of moisture condenser can be calculated as follows:

$$A_{air} \approx \dot{Q}_{moist,cond.} / \bar{U}_{air} \cdot LMTD \quad (7.17)$$

7.5 REFRIGERANT CONDENSERS

When the refrigerant leaves the heat pump compressor, it enters the condenser as a superheated vapor (state A) and leaves it as a subcooled liquid (state C) (Figure 17.6a). Refrigerant-to-air finned condensers reject heat from condensing refrigerants to outside drying air. Refrigerant flows through the tubes and a fan forces the drying air between the fins and over the tubes. The internal heat transfer of refrigerant condensers can be separated into three sections: superheated, saturated, and subcooled. The heat rejected from each section can be calculated by evaluating the refrigerant enthalpies at each section of inlet and outlet.

Overall, heat transfer coefficients of refrigerant condensers account for three thermal resistances: (i) forced boiling (vaporizing) thermal resistance on the refrigerant-side; (ii); conductive thermal resistance through the tube wall; and (iii) convective thermal resistance on the air-side:

$$\bar{U}_{refr} = 1 / \left(\frac{1}{\eta_{refr} \cdot h_{refr}} + \frac{\delta_{wall}}{k_{wall}} + \frac{1}{\eta_{air} \cdot h_{air}} \right) \quad (7.18)$$

where
$\bar{U}_{refr}$ is the overall heat transfer coefficient of refrigerant condenser ($W/m^2 \cdot K$);
η_{refr} is the refrigerant-side surface heat transfer efficiency (–);

h_{refr} is the refrigerant heat transfer coefficient ($W/m^2 \cdot K$);
δ_{wall} is tube wall thickness (m);
k_{wall} is the thermal conductivity of tube wall material (copper or aluminum) ($W/m \cdot K$);
η_{air} is the air-side surface heat transfer efficiency (–); and
h_{air} is the air-side heat transfer coefficient ($W/m^2 \cdot K$).

Air-side heat transfer efficiency is defined as follows:

$$\eta_{air} = 1 - \frac{A_{fin}}{A_{air}} \left(1 - \eta_{fin}\right) \tag{7.19}$$

where
η_{air} is the air-side overall surface heat transfer efficiency (–);
A_{fin} is the fin heat transfer area (m^2);
A_{air} is the total (fin plus exposed base) air-side surface heat transfer area (m^2); and
η_{fin} is the heat transfer efficiency of a single fin (–).

Thermal power of refrigerant condensers can be calculated from the refrigerant-side enthalpy changes and flow rate or from the air-side temperature change, mass flow rate, and specific heat (see Figure 6.5b and c in Chapter 6) as follows:

$$\dot{Q}_{refr,cond} = \bar{\eta}_{refr,cond} \cdot \dot{m}_{refr} \cdot \left(h_2 - h_3\right) = \dot{m}_{air} \cdot \bar{c}_{p,\,air} \cdot \left(T_d - T_{mix}\right) \tag{7.20}$$

where
$\dot{Q}_{refr,cond}$ is the thermal power of refrigerant condenser (kW);
$\bar{\eta}_{refr,cond}$ is the overall thermal efficiency of refrigerant condenser (–);
$\dot{m}_{refr}$ is the refrigerant mass flow rate (kg/s);
h_2 is the specific enthalpy of refrigerant superheated vapor entering the condenser (kJ/kg);
h_3 is the specific enthalpy of refrigerant subcooled liquid leaving the condenser (kJ/kg);
$\dot{m}_{air}$ is the air mass flow rate (kg/s);
$\bar{c}_{p,air}$ is the average isobaric specific heat of air ($kJ/kg \cdot K$);
T_d is the dry-bulb temperature of air leaving the condenser (°C); and
T_{mix} is the dry-bulb temperature of air entering the condenser (°C).

Thermal power of refrigerant condensers can be also written as follows:

$$\dot{Q}_{refr,cond} = \bar{U}_{air} \cdot A_{air} \cdot (LMTD) = \bar{U}_{air} \cdot A_{air} \cdot \left(\frac{T_{air,out} - T_{air,in}}{\ln \dfrac{T_{cond} - T_{air,in}}{T_{cond} - T_{air,out}}} \right) \tag{7.21}$$

where

$\dot{Q}_{refr,cond}$ is the thermal capacity (heat transfer rate) of refrigerant condenser (kW);

$\bar{U}_{air}$ is the condenser air-side overall heat transfer coefficient ($W/m^2 \cdot K$);

A_{air} is the condenser air-side heat transfer area (m^2);

$LMTD = \dfrac{\Delta T_{max} - \Delta T_{min}}{\ln \dfrac{\Delta T_{max}}{\Delta T_{min}}} = \dfrac{T_{air,out} - T_{air,in}}{\ln \dfrac{T_{cond} - T_{air,in}}{T_{cond} - T_{air,out}}}$ is the logarithmic mean temperature difference (°C) (see Figure 7.16b),

where

$\Delta T_{max} = T_{cond} - T_{air,in}$

$\Delta T_{max} = T_{cond} - T_{air,out}$

T_{cond} is the condensing temperature (°C)

7.6 ENERGY REQUIREMENTS FOR CONVENTIONAL HOT AIR CONVECTIVE BATCH DRYERS

The detailed knowledge of energy requirements for conventional batch dryers, as well as energy efficiency of the production systems, allows better managing the energy generation and distribution.

Conventional hot air convective batch dryer generally has energy efficiency below 50%, mainly because of the multiple thermal losses associated with it: (i) air venting; (ii) dryer structure; (iii) air leakages; (iv) heat generation devices; and (v) heat transportation networks.

In a simple, open hot air convective drying system (schematically shown in Figure 7.17a), the wet material enters the drying chamber at state *A* with total mass $\left(M_{dry}+M_{wet}\right)$, initial moisture content MC_{in}, and dry-bulb temperature T_A. After drying, the dried material leaves the drying chamber at state *B*, temperature T_B, mass M_{dry}, and final moisture content MC_{fin}.

On the other hand, the drying air (statel) with mass flow rate $\dot{m}_{air}$ is heated at constant absolute humidity ($\omega_1 = \omega_2$) (process $1-2$) which then enters the dryer as hot and dry air at state 2. Finally, the drying air absorbs moisture from the dried material at constant specific enthalpy (process $2-3$) (Figure 7.17b).

The thermal power supplied from outside to heat the drying air is as follows:

$$\dot{Q}_{12} = \dot{m}_{air} \cdot \left(h_2 - h_1\right) \tag{7.22}$$

where

$\dot{Q}_{12}$ is the thermal power supplied for heating the drying air (kW);

$\dot{m}_{air}$ is the mass flow rate of drying air (kg/s);

h_1 is the specific enthalpy of drying air at the inlet of air heater (kJ/kg); and

h_2 is the specific enthalpy of drying air at the outlet of air heater (kJ/kg).

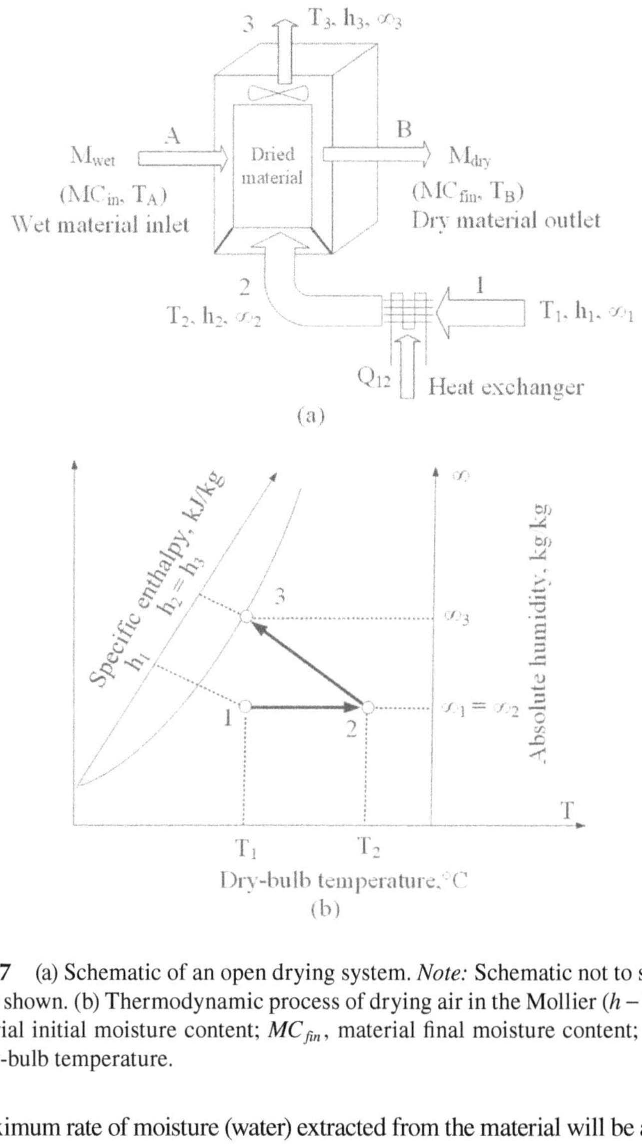

FIGURE 7.17 (a) Schematic of an open drying system. *Note:* Schematic not to scale; not all components shown. (b) Thermodynamic process of drying air in the Mollier ($h-T$) diagram. MC_{in}, material initial moisture content; MC_{fin}, material final moisture content; $\dot{Q}$, heat rate input; T, dry-bulb temperature.

The maximum rate of moisture (water) extracted from the material will be as follows:

$$\dot{M}_{moisture} = \dot{m}_{air} \cdot (\omega_3 - \omega_1) = \dot{m}_{air} \cdot (\omega_3 - \omega_2) \tag{7.23}$$

where

$\dot{M}_{moisture}$ is the maximum rate of moisture (water) extracted from the dried material (kg_{water}/s);

$\dot{m}_{air}$ is the mass flow rate of drying air (kg/s); and

ω_1, ω_2, and ω_3 are absolute humidity of the drying air at states 1, 2, and 3, respectively ($kg_{water}/kg_{dry\,air}$).

If $\overline{c}_p$ is the average specific heat at constant pressure of dried material, the thermal power balance of the simplified drying process represented in Figure 7.17b, will be as follows:

$$\dot{m}_{air} \cdot h_3 - \left(\dot{m}_{air} \cdot h_2 + \dot{M}_{wet} \cdot \overline{c}_{p,water} \cdot T_A\right) = \dot{M}_{dry} \cdot \overline{c}_{p,dry} \cdot (T_A - T_B) \tag{7.24}$$

where

$\dot{m}_{air}$ is the mass flow rate of drying air (kg/s);
h_3 is the specific enthalpy of drying air at drying enclosure exit (kJ/kg);
h_2 is the specific enthalpy of drying air at the drying enclosure inlet (kJ/kg);
$\dot{M}_{wet}$ is the mass flow rate of wet material at drying enclosure inlet (kg/s);
$\overline{c}_{p,water}$ is the average specific heat of water at drying enclosure inlet (kJ/kg·K);
T_A is the temperature of material at drying enclosure inlet (°C);
$\dot{M}_{dry}$ is the mass flow rate of dry material at drying enclosure exit (kg/s);
$\overline{c}_{p,dry}$ is the average specific heat of dry material at drying enclosure exit (kJ/kg·K); and
T_B is the temperature of dried material at drying enclosure exit (°C).

Figure 7.18 schematically represents the main energy (requirements) components of typical conventional, forced-air convection batch dryers.

External energy supplied to conventional hot air convective batch dryers should (i) move moisture inside the dried material and vaporize it from the material exposed surfaces; (ii) compensate air vent heat losses (Q_1); (iii) compensate air leakage heat losses (Q_2); (iv) warming-up dried material and dryer structure and equipment (Q_3); (v) compensate structure heat losses (walls, ceiling, doors, and floor) (Q_4); (vi) compensate heat losses by conduction along the supporting framework, and by convection and radiation from the dryer outer surfaces, especially in cold climates; (vii) compensate the heat losses of heat production devices and heat delivery systems; (viii) compensate heat losses during dryer shutdown and off-load periods; and (vii) energize the dryer overhead fans for air circulation (E_{fan}).

Based on previously identified components, the total energy requirements of conventional closed hot air convective batch dryers could be expressed as follows:

$$E^{conv}_{total,required} = \left(E_{fans} + E_{controls}\right) + \sum_{i=1}^{i=5} Q_i \tag{7.25}$$

where

$E^{conv}_{total,required}$ is the total (heat and electrical) energy required for conventional hot air convective dryers (kWh);
E_{fans} is the electrical energy consumed by overhead fans during the entire drying cycle (kWh);
$E_{controls}$ is the electrical energy consumed by dryer control devices during the entire drying cycle (kWh); and
$Q_i\left(i = 1\text{–}4\right)$ is the supplementary (and backup) (fossil and electrical) heat required during the entire drying cycle (kWh).

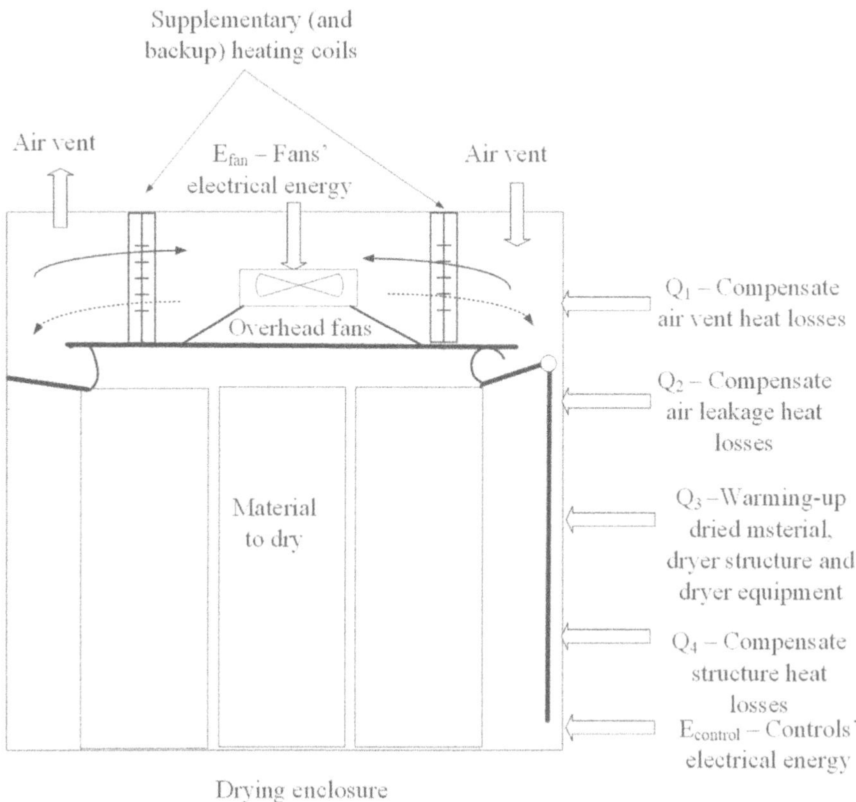

FIGURE 7.18 Energy components of typical conventional hot air convective batch dryers. *Note:* Schematic not to scale; not all components shown.

7.7 ENERGY REQUIREMENTS FOR HEAT PUMP-ASSISTED DRYERS

Total amount of energy requirements (consumed) by heat pump-assisted drying systems depends on many factors: (i) nature of the material to dry (including density, initial and final moisture contents, and actual drying schedule used); (ii) dryer type, size and construction materials, environmental conditions, type of heating, air temperature and velocity, venting, and control systems); (iii) average (estimated) drying time.

To the sensible and latent heat recovered by the heat pump from the wet material packages during drying, the equivalent electrical energy consumed by the heat pump compressor and the blower and dryer's fans must be added. In addition, for preheating the material stack(s), stickers and sleepers, the dryer envelope (structure, insulation, slab, foundation) and other auxiliary equipment, and to cover the dryer heat losses through walls and floor, supplementary (backup) heating energy is required for most of the heat pump-assisted dryers.

Figure 7.19 shows the main components of energy required by heat pump-assisted batch dryers equipped with remote condensers. Heat pump components (as compressors, evaporators, VSBs, and controls) are located inside separate mechanical rooms,

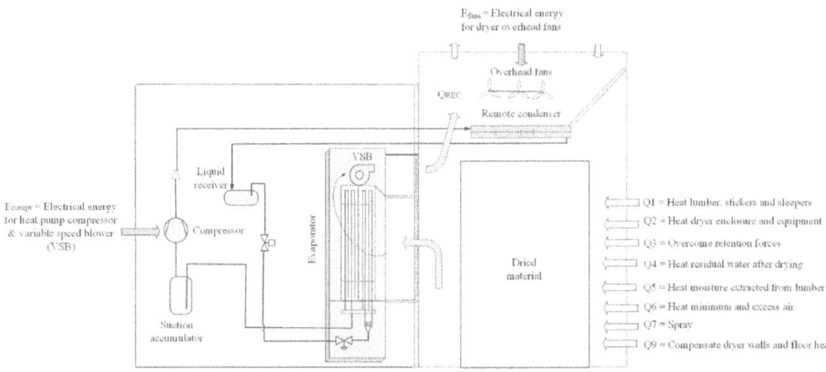

FIGURE 7.19 Energy requirements of typical heat pump-assisted lumber batch dryers using *split* mechanical vapor compression heat pumps and remote condensers. Similar energy requirements are for heat pump-assisted lumber batch dryers equipped with *compact* mechanical vapor compression heat pumps. *Note:* Schematic not to scale; not all components shown.

and remote condensers are installed inside the hot air convective drying enclosures. Supplementary (and backup) heat is supplied by electrical resistances and oil- or natural gas-burned boilers. Central overhead fans circulate the air through the stacks of the product to dry. Some of the drying chamber air vents (two or three) are kept active. They should open when the central (overhead) fan rotation change direction to avoid air implosion hazards, and also when the actual dry bulb temperature exceeds its set points.

The total energy required by heat pump-assisted lumber (wood) batch dryers can be calculated as follows:

$$E_{total,required}^{hp-assisted} = \left(E_{compressor} + E_{blower} + E_{fans} + E_{controls}\right) + \sum_{i=1}^{i=n} Q_i - Q_{REC} \tag{7.26}$$

where

$E_{total,required}^{hp-assisted}$ is the total energy (heat and electrical) required for a drying cycle (kWh);

$E_{compressor}$ is the electrical energy consumed by heat pump compressor during the entire drying cycle (kWh);

E_{blower} is the electrical energy consumed by heat pump drying air blower during the entire drying cycle (kWh);

E_{fans} is the electrical energy consumed by dryer overhead fans during the entire drying cycle (kWh);

$E_{controls}$ is the electrical energy consumed by dryer and heat pump control devices during the entire drying cycle (kWh);

Q_i is the supplementary (and backup) (fossil and/or electrical) heat required during the entire drying cycle (kWh) (see Figure 7.19); and

$Q_{REC} = Q_{REC}^{sensible} + Q_{REC}^{latent}$ is the total (sensible plus latent) heat recovered from the hot and moist drying air by air-to-air mechanical vapor compression heat pump (and returned to drying enclosure) during the entire drying cycle (kWh).

REFERENCES

Chua, K.J., S.K. Chou, J.C. Ho, N.A. Hawlader. 2002. Heat pump drying: Recent developments and future trends. *Drying Technology* 20(8):1579–1610.

Incropera, F.P., D.P. DeWitt. 2002. *Fundamentals of Heat and Mass Transfer*, 5th edition, John Willey & Sons, Inc., New York, NY.

Lewis, D. 2003. High temperature dehumidification systems. U.S. Patent 0208923.

McQuiston, F.C., J.D. Parker. 1988. *Heating, Ventilating, and Air Conditioning: Analysis and Design*, 3rd edition, John Wiley & Sons, New York, NY, 746 pp.

McQuiston, F.C., J. D. Parker, J. D. Spitler. 2005. *Heating, Ventilating, and Air Conditioning: Analysis and Design*, 6th edition, John Wiley & Sons, Inc., New York, NY.

Minea, V. 2006. Hardwood drying with low temperature heat pumps. In: *Proceedings of the 15th International Drying Symposium*, Volume C, pp. 1757–1762. Budapest, Hungary, August 20–23.

Minea, V. 2015. High-temperature heat pump-assisted softwood dryer: Sizing and control requirements & energy performances. In: *Proceedings of the 24th International Congress of Refrigeration (ICR2015)*. Yokohama, Japan, 16–22 August.

Minea, V. 2019. *Industrial Heat Pump-Assisted Wood Drying*. CRC Press, Boca Raton, FL.

Mujumdar, A.S., S. Devahastin. 2000. Fundamental principles of drying. In: *Mujumdar's Practical Guide to Industrial Drying: Principles, Equipment and New Developments*, edited by S. Devahastin, Thananuch Business Ltd. Publication, Bangkok, Thailand.

8 Efficient Operation and Control Rules

8.1 INTRODUCTION

After gaining knowledge of drying mechanisms of solid porous materials such as fruits, vegetables, and woods, the major challenge is attaining optimum design (size) of air-to-air mechanical vapor compression heat pumps coupled with hot air convective dryers, with efficient operation (and control) rules of dehumidification processes. System physical and thermodynamic coupling (integration), operating and control rules, as well as the usage of relevant drying schedules to save energy, reduce cost, and improve the quality of dried products are strongly interconnected.

Heat pump-assisted batch dryers are subject to special (particular) operation and control rules because of (Minea 2009a, 2009b) (i) refrigerant thermophysical-specific properties, (ii) location of heat pumps' condensers (outside or inside drying enclosures), and (iii) relatively high temperature differences existing between drying enclosures and heat pumps' mechanical rooms.

Mechanical components of compact-type and split-type low-, medium-, and high-temperature mechanical vapor compression heat pumps coupled with hot air convective batch dryers can be placed outside (Figure 8.1) or partially inside and partially outside the drying enclosures (see Figure 8.7), respectively. The heat pump basic components include (Figure 8.1) (i) refrigerant compressor, (ii) at least two heat exchangers (evaporator and condenser), (iii) refrigerant liquid receiver (LR), (iv) refrigerant liquid valve (LV), (v) adiabatic expansion valve (EXV), and (vi) motorized air dampers (MD).

Appropriate thermodynamic design, as well as the heat pump and dryer optimum coupling, running, and control rules must provide safe operation of low-, medium-, and high-temperature heat pump-assisted dryers. Such systems should maintain reasonably uniform drying conditions throughout the entire drying cycles. The scope is to remove enough quantities of moisture from dried products, condense it efficiently, and finally ensure good quality of dried materials.

On the other hand, wrong operation rules of mechanical vapor compression heat pumps coupled with hot air convective batch dryers could cause drawbacks (Minea 2010a, 2010b) (i) too low compressor suction pressures and temperatures; (ii) too high compressor discharge pressures; (iii) compressor frequent on/off cycling; (iv) excessive refrigeration migration during the preheating steps and heat pump standby periods; (v) moisture frosting on the evaporator external finned heat exchange surfaces; and (vi) eventual compressor mechanical failure, usury, severe damage, and shorter technical life.

To avoid such drawbacks, efficient operation rules should be ensured (Minea 2015): (i) quasi-continuous running of heat pumps' compressors by adequately

DOI: 10.1201/9781003618386-8

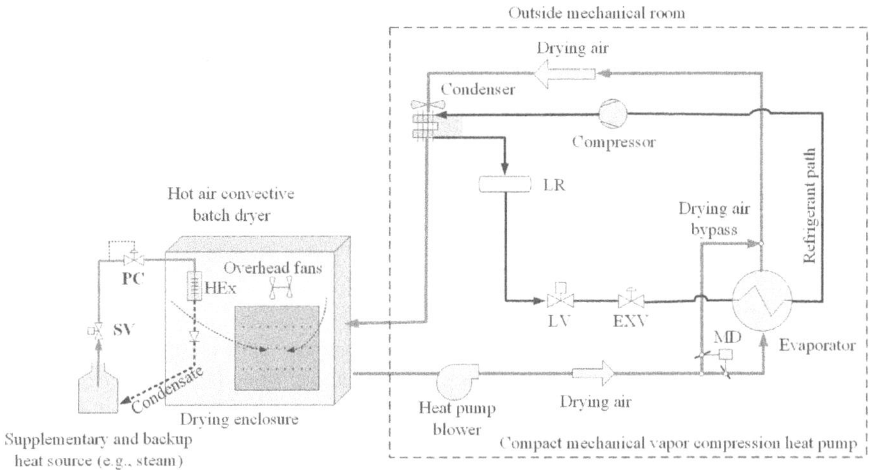

FIGURE 8.1 Hot air convective batch dryer coupled with a *compact* air-to-air subcritical mechanical vapor compression heat pump. EXV, expansion valve; HEx, steam-to-air supplementary (and backup) heat exchanger; LR, liquid receiver; LV, liquid valve; MD, motorized damper; PC, steam pressure regulator; SV, solenoid valve. *Note:* Schematic not to scale; not all components are shown.

correlating the refrigerant suction superheating to the refrigerant thermophysical properties in order to prevent saturated vapor or liquid droplets from entering the compressor suction port; (ii) setting the wet-bulb temperatures of drying air according to the actual measured wet-bulb temperatures to prevent the compressor frequent on/off cycling; (iii) managing the refrigerant migration by using liquid valves and/or artificially cooled suction accumulators (SAs) able to prevent refrigerant liquid from entering the compressors' suction lines; and (iv) in the case of large- and very-large-scale split-type heat pump-assisted dryers, improving supplementary (and backup) heating supply methods to prevent thermal shocks at the top of remote condensers and thus providing stable compressor discharge pressures and temperatures. In other words, to prevent any hazards or risks specific to heat pump-assisted dryers that can occur during drying, safe operating parameters and appropriate control of drying processes must be provided to the heat pump compressors and heat exchangers by adequate operation and control strategies (rules) (Minea 2010a, 2010b; Strumillo 2006).

8.2 QUASI-CONTINUOUS COMPRESSOR RUNNING

Inappropriate operation and control rules may result in low drying (dehumidification) efficiency of air-to-air mechanical vapor compression heat pumps coupled with hot air convective dryers. One of the most important dangers that could lead to risks such as compressor failures (mechanical damages) and technical life reduction may come from compression of saturated vapor of certain refrigerants if inadequate amounts of evaporator superheating are not ensured.

8.2.1 Set Evaporator Superheating

Among some simple but efficient methods to ensure safe operation of heat pump compressors is the control of the amount of evaporator refrigerant superheating (which practically determines the compressor suction temperature). It helps to avoid refrigerant saturated or vapor–liquid mixtures or even liquid drops to enter the compressor suction line. The control of evaporator refrigerant superheating should be adapted to the thermophysical properties of refrigerants to prevent damaging the reciprocating compressors.

8.2.1.1 Low-Temperature Refrigerants

Typically, the low-pressure (low-temperature) refrigerant vapor is superheated within the evaporator prior to entering the compressor suction line. The amount of superheating process *AB* (Figure 8.2) is controlled by the expansion valve according to the saturated evaporating pressure and temperature, and the actual compressor suction temperature. For most traditional low-temperature refrigerants, suction superheating of about 5°C is sufficient to prevent refrigerant saturated or wet vapor from entering the compressor at the end of adiabatic ($ds = 0$) compression process BC. This is possible due to the slope of the adiabatic compression curves in the superheated vapor zone of most low-temperature refrigerants (such as HFC-134a and HFC-410A) versus the position of their saturated vapor curves (Figure 8.2). In other words, if the compressor sucks up superheated vapor at state *B* before the adiabatic compression process BC, the isentropic curve does not touch or cross the refrigerant saturated vapor line.

8.2.1.2 High-Temperature Refrigerants

With certain high-temperature refrigerants (such as HFC-236fa and HFC-245fa), the evaporator superheating must be higher compared to those of most of conventional refrigerants.

As can be seen in Figure 8.3, with high-temperature refrigerants, the isentropic (adiabatic) curves in the superheated vapor zones are more vertically oriented compared to those of common low-pressure (low-temperature) refrigerants. Therefore, with high-temperature refrigerants, if the superheating amount is too small (e.g., 5°C)

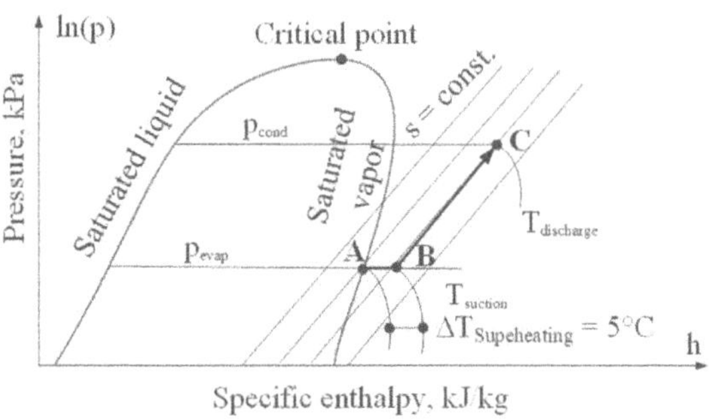

FIGURE 8.2 Usual superheating amount of most low-temperature conventional refrigerants. *cond.*, condensing; *evap*, evaporating; *p*, pressure; *s*, specific entropy; *T*, temperature.

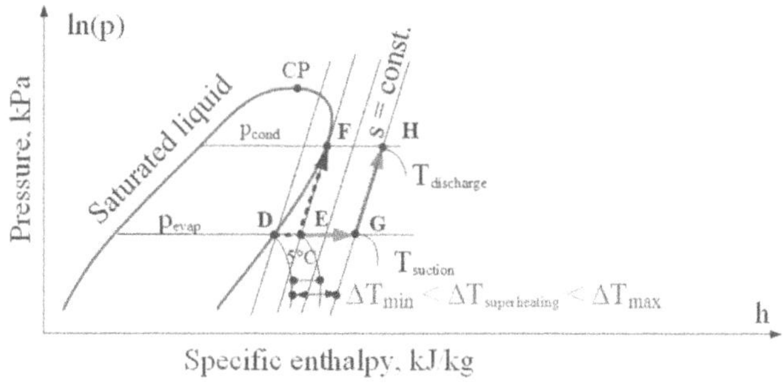

FIGURE 8.3 Optimum superheating amount for high-temperature refrigerants. *cond*., condensing; *evap*., evaporating; *p*, pressure; *s*, specific entropy; *T*, temperature.

(process *DE* in Figure 8.3), the compression adiabatic curve (*EF*) could touch or even cross the refrigerant saturation line. In this case, the thermodynamic state of the refrigerant leaving the compressor at state *F* can be located on the saturated line or even inside the saturated vapor–liquid zone. Thus, it may contain liquid droplets that could damage reciprocating compressors and/or reduce their useful technical life.

For high-pressure refrigerants, it was found that the optimum range of superheating $\Delta T_{superheating}$ (process *DG*) should be between a minimum value ΔT_{min} (e.g., 15°C) and maximum value ΔT_{max} (e.g., 25°C) at any saturated suction pressure (Minea 2010a, 2015):

$$\Delta T_{min} \lesssim \Delta T_{superheating} \lesssim \Delta T_{max} \tag{8.1}$$

where

ΔT_{min} is the minimum superheating amount (°C);
$\Delta T_{superheating}$ is the optimum superheating amount (°C); and
ΔT_{max} is the maximum superheating amount (°C).

To keep the evaporator superheating inside the desuperheating range (according to Equation 8.1), the following simple control strategy should be implemented by installing at the evaporator outlet (compressor inlet) a pressure ($p_{sat,evap}$) sensor and a compressor suction temperature reader ($T_{suction} = T_G$) (Figure 8.3). The pressure sensor ($p_{sat,evap}$) will send proportional signals to the heat pump process controller which will automatically calculate the actual saturated evaporating temperature ($T_{sat,evap}$) (ASHRAE 2013):

$$T_{sat,evap} = f\left(p_{sat,evap}\right) \tag{8.2}$$

where

$T_{sat,evap}$ is the saturated evaporation temperature (°C); and
$p_{sat,evap}$ is the saturated evaporation pressure (kPa).

For each pair of saturated evaporation temperature ($T_{sat,evap}$) and compressor suction temperature ($T_{suction} = T_G$), the amount of evaporator superheating will be calculated as follows:

$$\Delta T_{superheating} = T_G - T_{sat,evap} \tag{8.3}$$

where

$\Delta T_{superheating}$ is the optimum evaporator superheating amount (°C);
T_G is the compressor suction temperature (°C); and
$T_{sat,evap}$ is the evaporation saturated temperature (°C).

With the calculated amount of evaporator superheating, the heat pump process controller will automatically modulate the opening of the electronic expansion valve to keep the superheating inside the desired range according to Equation 8.3. If the actual superheating is, for example, 0.5°C below the lower (ΔT_{min}) limit, the expansion valve will slightly close to increase the evaporator superheating. If the actual superheating is, for example, 0.5°C higher than the upper limit (ΔT_{max}), the expansion valve will slightly open to decrease the evaporator superheating.

8.2.2 Set Wet-Bulb Temperature of Drying Air

A major issue for heat pump-assisted drying systems consists of appropriate setting of drying air wet-bulb temperature because the running (on/off) sequences of heat pumps' compressors strongly depend on it. Too short on/off cycles and many operating transient regimes contribute to lower the system overall drying efficiency and even to compressor shorter technical life. In addition, the drying air wet-bulb temperature is the principal parameter that determines when the heat pumps' compressors should start and when stop. To avoid the above-mentioned troubleshooting, an efficient method has been developed to set the dryer wet-bulb temperature automatically according to the actual measured values (Minea 2010a, 2010b).

During the warming-up step, the material to dry, dryer's structures, and auxiliary equipment of heat pump-assisted system are preheated, but heat pump compressor is not running. The heat pump starts at the end of the preheating period when the wet-bulb temperature of the drying air inside the drying enclosure reaches its first preset values (let say, $WBT-1$) (Figure 8.4). At that moment, the actual wet-bulb temperature of drying air is higher than the first preset value $WBT-1$. Because the compressor is now running and the dryer is a relatively tight enclosure with all air vents closed, the difference between the actual air wet-bulb temperature and the first preset air wet-bulb temperature ($WBT-1$) gradually decreases over the time.

When the difference between the actual (measured) wet-bulb temperature and the first preset value ($WBT-1$) reaches, for example, 5°C, the process controller automatically lowers the previous preset value $(WBT-1)$ to the second set value of

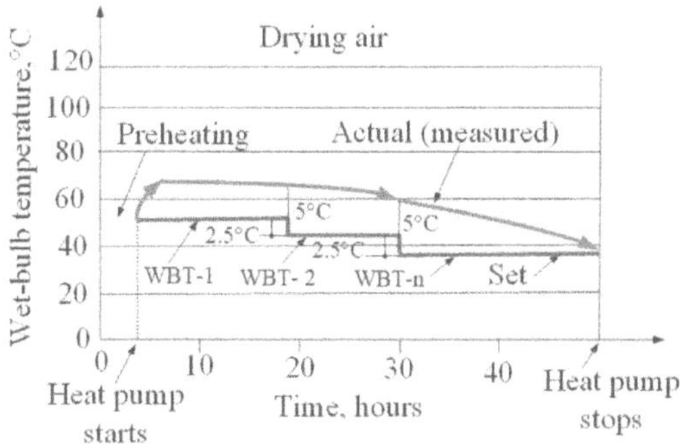

FIGURE 8.4 Actual (set and actual/measured) wet-bulb temperatures of drying air during drying cycles. *Note:* Graph not to scale. *WBT*, wet-bulb temperature.

drying air wet-bulb temperature ($WBT-2$), for instance, by 2.5°C, according to the following equation (Minea 2011):

$$WBT_{set}^{n} = WBT_{set}^{n-1} - 2.5°C \tag{8.4}$$

where
- WBT_{set}^{n} is the current set wet-bulb temperature of drying air (°C);
- WBT_{set}^{n-1} is the previous set wet-bulb temperature of drying air (°C);
- n is the current setting step of drying air wet-bulb temperature ($n = 1, 2, 3\ldots$) (–)

When the difference between the actual (measured) drying air wet-bulb temperature and the second ($WBT-2$) set drying air wet-bulb temperature reaches 5°C once again, the system process controller again automatically lowers the second set air wet-bulb temperature by 2.5°C (to $WBT-3$), and so on.

Toward the end of drying cycle, when the set air wet-bulb temperature ($WBT-n$) is, for instance, 1°C below the previous actual wet-bulb temperature value (WBT_{set}^{n-1}), the compressor will shut down automatically. The compressor will restart only if the actual air wet-bulb temperature becomes 1°C higher than the last set value of the air wet-bulb temperature. If the heat pump compressor starts and shuts down another three consecutive times during, for example, 10–20 minutes, the system process controller will stop the compressor and the drying cycle in dehumidification mode will end for good.

This developed control strategy (Minea 2011), which automatically lowers the air set wet-bulb temperature according to the temperature difference between the actual (measured) and the previous drying air set wet-bulb temperatures, prevents frequent on/off cycles of heat pump compressors and, thus, contributes to improve their reliability and technical life. That is because frequent and/or too short on/off running cycles involve too many transient thermodynamic regimes that may contribute to lower the system overall drying efficiency and to compressors wear.

8.2.3 Set Drying Air Temperature Depression

Prior to starting a drying cycle, drying air dry-bulb and wet-bulb temperatures are preset for each drying step. Both these temperatures are then continuously measured and their values automatically transferred to the process drying controller.

The strategy aiming to automatically set the wet-bulb temperature of drying air according to the actual measured values (see Section 8.2.2) should be coordinated with the actual drying air temperature depression, defined as the difference between drying air dry-bulb and wet-bulb temperatures (also called wet-bulb temperature depression) (Figure 8.5). As can be seen in Figure 8.5, at the beginning of any drying cycle, the drying air temperature depression is relatively low (e.g., between 3°C and 5°C). As the drying process progresses, the rate of moisture extracted from the dried material tends to diminish and, thus, the drying air temperature depression should increase progressively (e.g., up to 20°C) toward the end of the drying cycle.

By appropriately setting and controlling the evaporator superheating amount (Section 8.2.1), drying air temperature depression (Section 8.2.3) and efficiently controlling the refrigerant migration (see Section 8.2.4), heat pump-assisted dryers could be able to (practically) continually operate without frequent on/off cycles.

8.2.4 Efficient Management of Refrigerant Migration

Efficient management of refrigerant migration throughout the heat pump piping and components could be provided by (Minea 2010a, 2010b) (i) optimum (adequate) quantity (mass) of heat pump refrigerant charge; (ii) using refrigerant liquid valves; and (ii) using both refrigerant liquid valves and (modified) artificially cooled suction refrigerant accumulators, especially in the case of large and very large industrial-scale heat pump-assisted dryers.

8.2.4.1 Optimum Refrigerant (Mass) Charge

For all refrigeration and heat pump systems, the refrigerant charge (mass) is a critical parameter. Most of the refrigerant charge is found within the condensers and

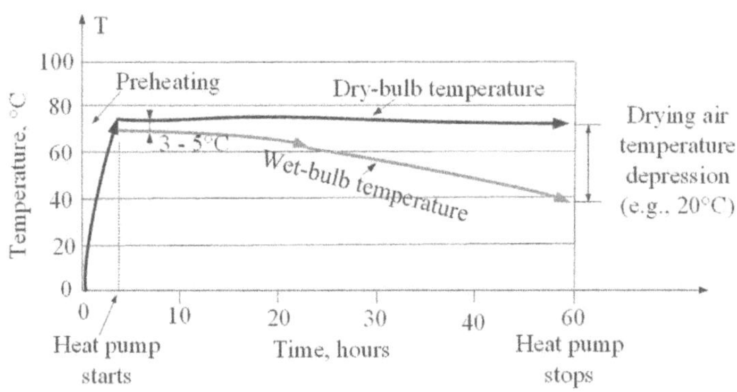

FIGURE 8.5 Example of optimum setting of drying air temperatures depression. *Note:* Graph not to scale.

evaporators, as well as inside the liquid receivers, liquid suction piping, and compressor SAs. Proper refrigerant charging contributes to normal energy efficiency of heat pump-assisted dryers, while improperly charged systems provide poor heat transfer coefficients and reduced energy efficiency. For example, a 15% under-charge can result in a 25% decrease in heat pump coefficient of performance. As the charge drops further, the heat pump efficiency may decrease up to 50% or more. Moreover, a poorly charged heat pump will not properly cool or lubricate the compressors and, thus, may significantly shorten their technical life. Similar losses of efficiency result from overcharging the heat pumps. Either overcharging or undercharging finally increase the operating costs and maintenance expenses.

Correct refrigerant charge could be achieved by comparing the measured temperature of subcooled liquid refrigerant at the condenser outlet with the actual condensing temperature (determined as a function of actual measured condensing pressure). This temperature difference should be in the range of 5–7°C.

8.2.4.2 Pump-Down with Liquid Valve

In *compact* (and also *split*) heat pump-assisted dryers, during the preheating (warming-up) steps, heat pump stand-by and off periods, the refrigerant naturally migrates toward the compressor and the SA that are large metallic components with relatively high thermal inertia (Figure 8.6). The temperatures of metallic SAs and compressors is generally lower than those of the rest of heat pumps' components. Therefore, the refrigerant migrates and condenses inside them. When the compressor is called to start, it will first try to pump the liquid refrigerant from its crankcase

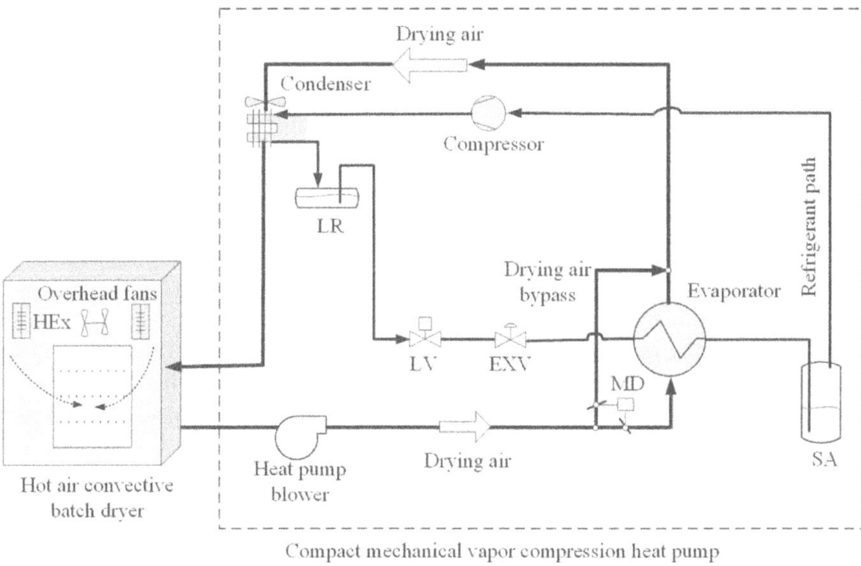

FIGURE 8.6 Manage refrigerant migration with liquid valve. EXV, expansion valve; HEx, supplementary (and backup) heat exchanger; LR, liquid receiver; LV, liquid valve; MD, motorized damper; SA, suction accumulator.

and SA. Because the liquid is incompressible, the pressure differential will be too low, and thus the compressor automatically shuts down. After a short period of time, the compressor starts up again, and several such on/off cycles may occur until the whole quantity of liquid is pumped out and the set pressure differential is restored. Such too frequent and too short on/off cycles reduce the useful life of reciprocating compressors and can damage their moving parts (pistons, connecting rods) and cylinders (Minea 2015).

To prevent excessive refrigerant migration and liquid from entering the compressor suction line, an efficient method consists of using a liquid valve (LV) between the condenser and the liquid receiver (LR), as shown in Figure 8.6.

The liquid valve (LV) allows controlling the refrigerant migration during the compressor off and standby periods. When the compressor shuts down, the liquid valve (LV) immediately closes, but the compressor continues to run for additional (let say) 10–15 seconds. The refrigerant from the evaporator EV is, thus, pumped down and stored inside the condenser and the liquid receiver (LR). At the end of this short running sequence, the evaporator is practically free of refrigerant, and the condenser and liquid receiver store major part of the refrigerant charge. When the compressor is due to restart, the liquid valve (LV) first opens for approximately 5–8 seconds, while the compressor is still off. During this period of time, a certain amount of liquid flows from the condenser to the inlet port of the expansion valve, and then the compressor restarts. At this moment, the oil pressure difference will be higher than the minimum differential limit, and the compressor will not shut down. Such a simple control rule allows pumping down the refrigerant prior to each compressor shutdown which prevents the liquid to enter the compressor suction port at each starting sequence and, thus, to avoid damaging it.

In other words, the liquid valve (LV) aims to prevent the refrigerant to migrate and store inside the compressor crankcases and/or inside the SA during off and stand-by periods. Consequently, when the compressor is called to restart, any or a very small quantity of refrigerant in liquid state will enter the compressor suction port.

8.2.4.3 Modified Suction Refrigerant Accumulator

In relatively small-scale heat pump-assisted dryers, the pump-down strategy using liquid valves could totally address the problem of the liquid from entering the compressor suction line (Section 8.2.4.2).

However, in the case of large- and very-large-scale (i.e., with compressor electrical power inputs higher than, let say, 100 kW) medium- and high-temperature *split*-type (Figures 8.7–8.9) heat pump-assisted dryers, where there is significant temperature differences (up to 100°C) between drying enclosures and mechanical rooms, the liquid valve control may not be entirely safe for the compressor mechanical integrity. When such heat pump-assisted drying systems are not running for relatively long periods of time, the refrigerant flows in both directions through (usually) leaking check valves *CV*1 even when the temperature difference between the air inside the dryer enclosure and the mechanical room becomes relatively small. During the preheating step, the heat pump does not run, and the refrigerant migrates from the drying enclosure (being at much higher temperature) to the SA and the compressor crankcase (which are metallic masses with high thermal inertia, and much lower temperatures where it condenses). When the compressor starts at the first drying step, it will suck up the stored liquid refrigerant stored inside its crankcase and SA.

To avoid such an undesirable situation, in large- and very large-scale heat pump-assisted drying systems, one of the following developed technical solutions could be used: (i) modified SAs coupled with small air-cooled condensing units having small evaporators immersed or coiled around the external surfaces of heat pump refrigerant accumulators; or (ii) non-modified refrigerant SAs totally immersed inside cold water tanks.

To reinforce the control of refrigerant migration, the modified suction accumulator ($MSA-1$) may be coupled to a small air-cooled condensing unit composed of a small evaporator (EV) installed inside $MSA-1$, a compressor (C), a condenser (CD), and an expansion valve (EXV), as shown in Figure 8.7. Few minutes before the heat pump compressor starts, the small condensing unit starts and cools the refrigerant liquid eventually stored inside $MSA-1$. Thus, the refrigerant liquid stored (e.g., inside the compressor crankcase and the heat pump evaporator) will migrate toward the colder $MSA-1$. Then, the small solenoid valve SV opens, the small refrigerant liquid pump LP starts up and the refrigerant liquid is pumped toward the heat pump liquid receiver LR via check valve $CV2$. After having pumped almost whole quantity of refrigerant liquid, the solenoid valve SV closes, the liquid pump LP shuts down, and the heat pump compressor starts up safely.

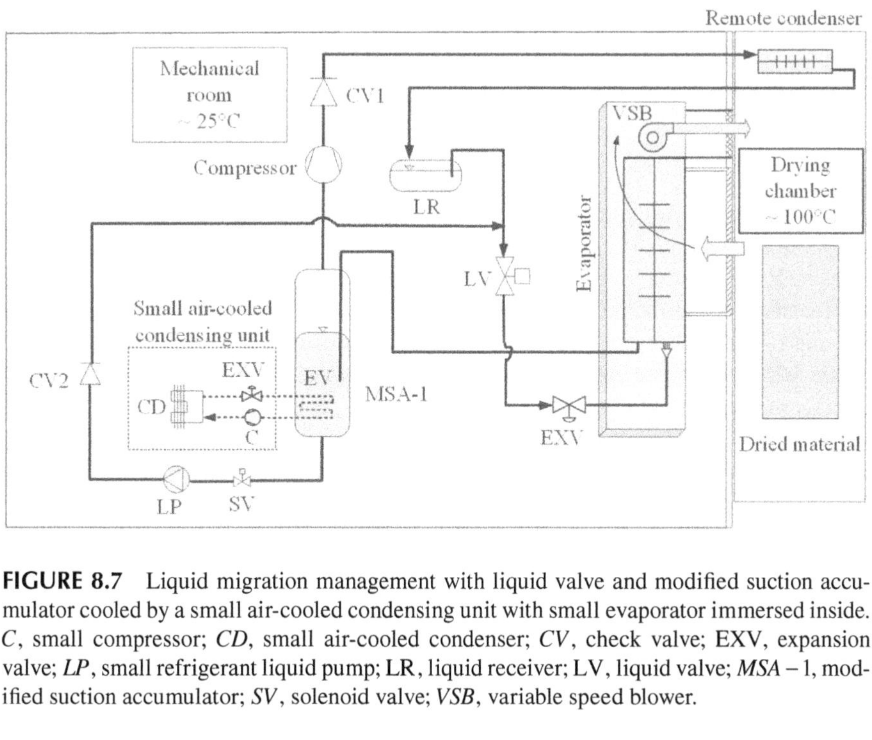

FIGURE 8.7 Liquid migration management with liquid valve and modified suction accumulator cooled by a small air-cooled condensing unit with small evaporator immersed inside. C, small compressor; CD, small air-cooled condenser; CV, check valve; EXV, expansion valve; LP, small refrigerant liquid pump; LR, liquid receiver; LV, liquid valve; $MSA-1$, modified suction accumulator; SV, solenoid valve; VSB, variable speed blower.

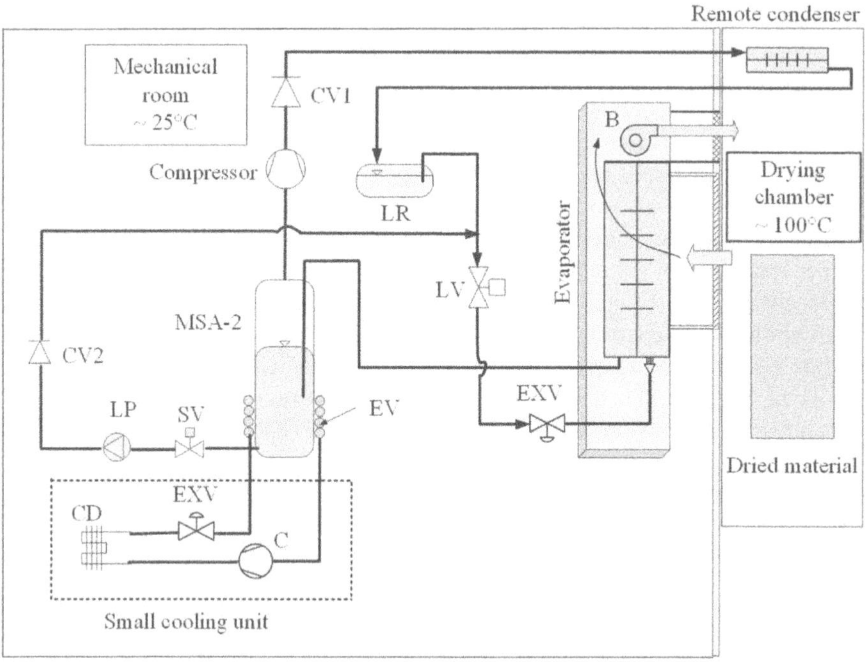

FIGURE 8.8 Liquid migration management with liquid valve and modified suction accumulator ($MSA-2$) cooled by a small air-cooled condensing unit with small external evaporator coiled around the external surface. *C*, compressor; *CD*, air-cooled condenser; *CV*, check valve; EXV, expansion valve; *EV*, evaporator; *LP*, refrigerant liquid pump; LR, liquid receiver; LV, liquid valve; $MSA-2$, modified suction accumulator; *SV*, solenoid valve; *VSB*, variable speed blower.

The SA could also be coupled with a small air-cooled condensing unit of which evaporator *EV* is attached around its external metallic surface ($MSA-2$) (Figure 8.8). The small compressor *C*, the air-cooled condenser *CD*, and the expansion valve EXV are located outside the modified suction accumulator $MSA-2$. Few minutes before the end of the preheating step, the small compressor *C* starts and the temperature inside *MSA-2* becomes lower. Therefore, the refrigerant stored inside the compressor crankcase, heat pump evaporator, and refrigerant piping migrates toward the heat pump modified suction accumulator $MSA-2$. After this relatively short sequence, the small solenoid valve *SV* opens and the small liquid pump *LP* starts up. The refrigerant liquid is thus pumped toward the liquid receiver LR via check valve $CV2$. After having pumped almost whole quantity of refrigerant liquid from the modified suction accumulator $MSA-2$, solenoid valve *SV* closes, liquid pump *LP* shuts down, and the heat pump compressor can start up safely (Minea 2010a, 2010b).

A third solution consists of inserting the heat pump SA inside a thermally insulated water tank (Figure 8.9). Cold water from the city network, a lake, or a river can be continuously circulated through the water receiver during short periods of time prior to starting the heat pump compressor. The purpose is to cool the external

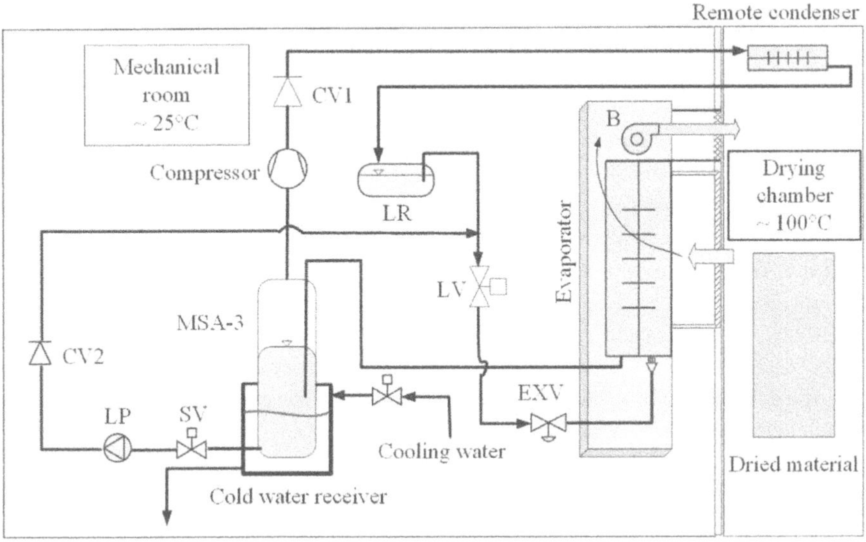

FIGURE 8.9 Liquid migration management with liquid valve and nonmodified suction accumulator ($MSA-3$) immerged in a cold water tank. C, compressor; CD, air-cooled condenser; CV, check valve; EXV, expansion valve; EV, evaporator; LP, refrigerant liquid pump; LR, liquid receiver; LV, liquid valve; $MSA-3$, modified suction accumulator; SV, solenoid valve; VSB, variable speed blower.

surface of nonmodified suction accumulator $MSA-3$. After having cooled the heat pump suction accumulator $MSA-3$, the cooling water returns to city sewer, lake, or river. Between the nonmodified suction accumulator $MSA-3$ and the heat pump liquid refrigerant line, an additional cooling circuit includes a small solenoid valve SV, a small liquid pump LP, and refrigerant piping with check valve $CV2$. At the end of preheating step, few minutes prior to the heat pump compressor starts, the cooling water enters the water receiver at temperatures between 5°C (in the winter) and 20°C (in the summer). Thus, the metallic envelope of nonmodified suction accumulator $MSA-3$ is cooled and the refrigerant migrates from the compressor crankcase, heat pump evaporator, and refrigerant piping toward the nonmodified suction accumulator $MSA-3$. Then, the small motorized valve SV opens, and the small refrigerant pump LP starts providing enough head pressure to pump almost whole quantity of liquid stored inside the nonmodified suction accumulator $MSA-3$ to the heat pump liquid receiver LR.

8.2.5 Avoid Thermal Shocks

In conventional hot air convective batch dryers (without heat pumps), heating agents (steam or hot water) are usually supplied from oil-burned or electric-powered boilers to heat exchangers (coils) installed inside the drying enclosures. In the case of steam used as a heating agent, when heat is required for drying enclosure, a large-capacity solenoid valve $L-SV$ fully opens and relatively large amounts of heat are supplied (see Figure 8.10). The solenoid valve $L-SV$ fully opens and closes according to the air dry-bulb temperature required inside the drying enclosure. When the solenoid valve $L-SV$ is fully open, the steam circulates at high flow mass rates through the heating coils located inside the dryer chamber.

In *split*-type heat pump-assisted dryers equipped with remote refrigerant condensers (Figure 8.10a) in heat pump-assisted drying systems, additional (supplementary) heat is needed mainly to (i) cover the whole dryer energy requirements, structure heat losses, and drying air leakages and other uncontrollable thermal losses; and (ii) compensate for the lack of heat pump heating capacities.

In heat pump-assisted (advanced) drying systems, particular control strategies are required because (Minea 2015) (i) remote condenser(s) are installed inside the drying enclosures; (ii) relatively high temperature differences exist between dryer enclosures and mechanical rooms of mechanical vapor compression heat pumps; and (iii) multi-blade overhead dryer fans circulate the drying air over both heat pump remote condenser(s) and material stack(s).

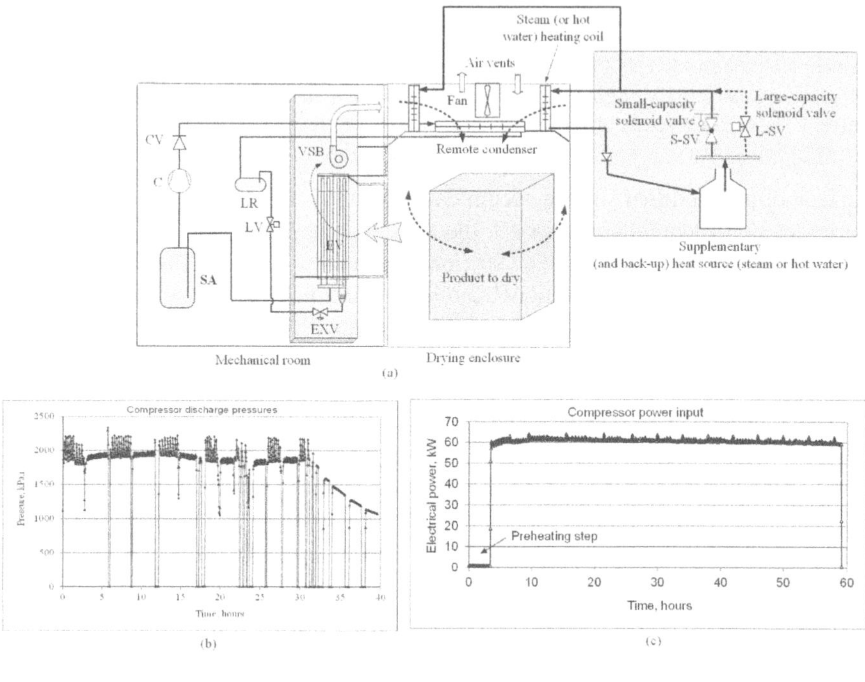

FIGURE 8.10 (a) Appropriate supplementary (and backup) steam (or hot water) supply with small-capacity modulating solenoid valve $S-SV$. (b) Example of abnormal operation of heat pump compressor if steam is supplied via large-capacity solenoid valve $L-SV$. (c) Example of heat pump compressor normal and continuous operation if steam is supplied via the small-capacity modulating valve $S-SV$. *C*, compressor; *CV*, check valve; *EV*, evaporator; EXV, expansion valve; LR, liquid receiver; LV, liquid valve; *SA*, suction accumulator; *LSV*, large-capacity solenoid valve.

8.2.5.1 Heat Supply during Warming-Up (Preheating) Step

During the preheating step of *split*-type heat pump-assisted dryers, the amount of heat transferred to the dryer enclosure by steam-to-air (or hot water-to-air) heat exchangers are much higher compared with the additional quantity of heat required when the heat pump compressor is running. Therefore, during the preheating steps, when the heat pump compressors are not running, the large-capacity solenoid valve $L-SV$ opens, and steam (or hot water) is safely directed at full flow rates through the distribution piping toward the dryer heating coils (Figure 8.10a).

8.2.5.2 Heat Supply during Heat Pump-Assisted Drying Steps

In the case of *split*-type heat pump-assisted dryers, heating exchangers are located at upper sides of drying enclosures, close to the heat pump remote condensers. When the dryer fans turn, the dryer warm and humid air first passes over the steam (or hot water) heat exchangers where it is heated, and then passes over the heat pump condenser(s) (Figure 8.10a).

Since the heat pump condensers are located relatively close to the heating coils, the supplementary heat supply via large-capacity solenoid valves $(L-SV)$ is not allowed because it could provide sudden and excessive thermal shocks to the heat pump remote condensers. If the remote condensers of split heat pump-assisted dryers are not adequately protected against such thermal shocks (e.g., with thermal screens, bafflers, etc.), they may induce excessive high discharge pressures and temperatures to heat pumps' compressor. During short periods of time, these temperatures could be much higher than the normal air temperatures entering the condenser. It may generate undesirable thermal shocks at the heat pump's condenser inlet and excessively increases the compressor discharge pressure and temperature. Consequently, exceeding the discharge pressure limits determine the compressor to frequently shut down and, thus, run with relatively short on/off cycles, as shown in Figure 8.10b. Such abnormal parameters may exceed the compressor safe discharge pressures (and temperature) operation limits, contribute to refrigerant and lubricating oil chemical degradation, and even to the compressor break down and/or premature wear.

To avoid such abnormal operation conditions, a small-capacity, high-precision modulating solenoid valve $(S-SV)$ must be installed in parallel with the large-capacity solenoid valve $L-SV$. In the case of steam, such a small-capacity valve will be activated by a pressure sensor to modulate the steam flow rate according to the steam leaving pressure. The steam flow rate should be adjusted to keep the outlet pressure in a range of approximately $5–10\%$ of the inlet steam pressure. With smaller and continuous relatively constant steam flow rates, no excessive thermal shock will be provided at the heat pump remote condenser, and the heat pump compressor will operate at relatively constant discharge pressures and temperatures. In this way, as shown in Figure 8.10c, the heat pump thermodynamic cycle will be stable and sudden thermal shocks at the inlet of remote condensers will be avoided (Minea 2015).

REFERENCES

ASHRAE. 2013. Fundamentals Handbook, SI edition, American Society of Heating, Refrigerating and Air Conditioning Engineers, Atlanta, GA.

Minea, V. 2009a. Improvements of compact and split high-temperature drying heat pumps. In: *Proceedings of the 8th World Congress of Chemical Engineering and Inter-American Drying Conference (IADC'09)*. Montréal, Canada, August 23–27.

Minea, V. 2009b. Avoiding failures of large-scale high-temperature drying heat pumps. In: *Proceedings of the 6th Asia-Pacific Drying Conference (ADC2009)*, pp. 179–187. Bangkok, Thailand, October 19–21.

Minea, V. 2010a. Improvements of high-temperature drying heat pumps. *International Journal of Refrigeration* 33(1):180–195.

Minea, V. 2010b. Improvements of high-temperature drying heat pumps. In: *Proceeding of International Refrigeration and Air Conditioning Conference at Purdue*, West Lafayette, IN, July 12–15.

Minea, V. 2011. Industrial drying heat pumps. In: *Mechanical Engineering Theory and Applications: Refrigeration—Theory, Technology and Applications*, edited by Mikkel E. Larsen, Nova Science Publishers, Inc., New York, pp. 1–70.

Minea, V. 2015. High-temperature heat pump-assisted softwood dryer: Sizing and control requirements & energy performance. In *Proceedings of the 24th IIR International Congress of Refrigeration, Improving Quality of Life, Preserving the Earth*. Yokohama, Japan, August 16–22.

Strumillo, C. 2006. Perspectives on development in drying. *Drying Technology* 24:1059–1068.

9 Conclusions

9.1 INTRODUCTION

Drying is one of the most energy-intensive operations in many industrial sectors (Lewis 1921; Sherwood 1929), where energy-saving and the need for lower CO_2 emission have become the top priority. Over the years, academia researchers and industrial engineers have employed various strategies to improve drying efficiency and product final quality in the context of increasing energy costs.

Based on conventional drying methods, new drying technologies, such as heat pump-assisted drying, have been developed to ensure energy performances similar to or higher than those of conventional drying methods while generating energy-savings and enhancing the final quality of dried products such as fruits, vegetables, and woods. However, despite promising features of heat pump-assisted dryers, the number of industrial-scale implementation today is not as many as expected.

This book points out that over the past few decades, materials such as fruits, vegetables, and woods have been dried with heat pump-assisted dryers, but relatively few published R&D studies have provided complete and/or pertinent information about initial input theoretical/experimental data, system size, running and control methods used, and/or reliable results. To avoid malfunctions and/or failures of mechanical vapor compression heat pumps coupled with hot air convective batch drying enclosures, as well as poor drying performances, the book outlines methods to avoid undesirable incidents and/or troubleshooting (such as uncontrollable refrigerant migration, inefficient supplementary heat supply, thermal shocks, and, finally, heat pump compressor mechanical failures). In the future, academia drying researchers and industry technicians should avoid potential mistakes and/or inadequate operation strategies (rules) likely to lead to system malfunction, equipment failures, and/or false drying performances.

The final objective of the book is to contribute to improving the design and operation rules of heat pump-assisted dryers and, thus, increase the number of their industrial implementations.

9.2 PRESENT SITUATION

Selection, design, and operation of heat pump-assisted batch dryers, where air-to-air mechanical vapor compression heat pumps coupled with hot air convective batch dryers are *compact* or *split* devices, depends on physical/chemical properties and drying characteristics of dried products.

In the case of *compact* heat pumps, the evaporators and condensers are installed in separate mechanical rooms adjacent to drying enclosures. The products are placed inside drying chambers where the drying medium (usually, air) flows parallel or perpendicular to the product surface.

DOI: 10.1201/9781003618386-9

In the case of *split* heat pumps, there are two chambers thermally isolated from one another: (i) drying enclosures with remote condensers; and (ii) mechanical rooms with the rest of the heat pump components (compressor, evaporator, and blower), air ducts, and controls; all temperature-sensitive components of *split* heat pumps are, thus, thermally separated from the hot drying chambers.

During the analysis of present situation regarding the pertinence and the reliability of heat pump-assisted drying concepts, as well as of reported theoretical and/or experimental results over the past few decades by several published R&D works, the following observations were made (not listed in order of importance):

a. Single stage air-to-air mechanical vapor compression heat pumps coupled with hot air convective batch drying enclosures are complex systems because of interdependency of many thermodynamic parameters.
b. Inadequate, unnecessarily complicated, and/or unworkable concepts have often been proposed (developed); or, increasing the complexity of coupled (integrated) thermodynamic systems cannot contribute to further industrial deployments, despite their technical, economic, and environmental advantages; unfeasible or doubtful (so-called *improved*) concepts can disappoint both young academia researchers and industry technicians who may not trust the new proposed concepts and may compromise future technical advancements and industrial applications.
c. Incomplete or even wrong theoretical and/or experimental results, including energy (drying) performances, have sometimes been published.
d. Several published journal articles and conference papers have reported relatively high energy performances, but have not provided essential data (information), for example: (i) material input/output quantities (masses, volumes), (ii) material initial/final moisture contents, (iii) heat pump dehumidification capacity (evaporator thermal power), (iv) compressor-rated (nominal) electrical input power, (v) condenser heating and/or heat delivery capacity, and (vi) variations of compressor discharge pressures and temperatures throughout the entire drying cycles.
e. Drying schedules were not always provided in terms of set dry- and wet-bulb temperatures, or temperature depression in relation to the drying air relative and absolute humidity and flow rates.
f. Drying curves of dried products were sometimes provided without specifying whether the drying air moisture contents were measured and how (oven, continuously or intermittently).
g. Preheating and supplementary (backup) heating requirements and specifications about heat source (electrical or fossil) employed were provided in very few studies.
h. Methods aiming to match the heat pump dehumidification capacity (i.e., heat pump evaporator thermal power) with the product thermal dewatering rate and/or to control the temperature and relative humidity of the drying air were rarely employed, while external heat rejection devices were frequently employed.

i. Usage of same drying air flow rates over both heat pump evaporator and condenser in many experimental setups resulted in wrong reported energy performances.
j. Some published studies wrongly claimed that heat pumps having given nominal thermal (dehumidification) capacities or heat pump electrical rated power inputs could efficiently dry different products (solids, liquids), in different drying modes (batch, continuous, or intermittent), with different drying mediums (air, inert gases, CO_2), and at negative or positive temperatures, without modifying and/or adjusting the dryer-heat pump coupling and/or control strategies.
k. However, contrary to such statements, without substantial structural modifications of coupled thermodynamic systems and/or of drying schedules and control rules, a given heat pump-assisted drying setup cannot be used efficiently (i) for drying all kinds of products, in any physical state, such as solid, liquid or semi-liquid, agro-food, heat-sensitive, biological products, and wood; (ii) in any drying mode (batch, conveyor, continuous, intermittent); (iii) over large temperature ranges (e.g., from –30°C to 110°C); (iv) with any drying medium (air, oxygen, inert gases, CO_2, etc.); and (v) by using the same refrigerant, without modifying the refrigerating configurations, design of heat exchangers, and/or system control rules.
l. Potential heat pump malfunctions, such as the following, were practically never reported: (i) frequent and too short compressor on/off cycles; (ii) too high or too low refrigerant pressures and temperatures; and (iii) evaporator frosting.

9.3 FUTURE R&D NEEDS

Many emerging drying technologies that aim to replace conventional ones still require significant R&D efforts "to be marketable" in the future (Strumillo et al. 1995; Mujumdar and Huang 2007; Hosseinpour and Martynenko 2023; Martynenko and Mujumdar 2024; Perré 2024).

Increased requirements on process efficiency and product quality pose new challenges for the drying industry, and demonstrate a growing need for optimization of control strategies (Winiczenko et al. 2018).

Because the main scope of R&D work is to provide industry with workable solutions, the future R&D studies should make heat pump-assisted drying systems reliable for each specific dried product. The true challenge is to make these systems feasible by providing relevant information about systems' integration, schedules, control strategies, operating rules, and energy performances. Obviously, without improving the coupling of batch dryers with air-to-air mechanical vapor compression heat pumps and their coupled operation throughout the whole drying process, no useful technological advancements can be achieved. The lack of reliable coupled systems may compromise the future applicability of standard and/or innovative heat pump-assisted drying concepts. In other words, scientific studies must not unnecessarily complicate the proposed technologies with unworkable solutions.

This book suggests that future theoretical and experimental R&D works, as well as journal articles and conference papers on heat pump-assisted drying concepts, should include the following (not listed in order of importance):

- Optimum configurations coupling hot air convective batch dryers with air-to-air nonreversible mechanical vapor compression heat pumps.
- Strategies such as drying air evaporator bypasses or variable speed blowers, variable speed or multiple compressors, or intermittent drying to continuously match materials' dewatering rates with heat pumps' actual dehumidification capacities.
- Optimum design (size) of heat pumps' nominal dehumidification capacities and efficient running and control rules.

Future published R&D works should also perform the following (not listed in order of importance):

1. Focus on the simplicity of concepts to increase the reliability and dehumidification efficiency of industrial drying applications.
2. Describe the specific drying mechanism of each dried porous solid material, such as fruits, vegetables, and woods.
3. Understand and describe the role of materials' micro- and macro-structures in heat and moisture transfer mechanisms.
4. To solve coupled and nonlinear equations, and to model (at microscopic and macroscopic scales) the drying processes, advanced skills in applied and computational mathematics are required.
5. Provide pertinent information about dryer-heat pump coupling (integration), design and control rules, operating and control rules, as well as about actual operating parameters and predicted and/or achieved energy performances; inadequate integration and wrong operation parameters may provide troubles as compressor high discharge pressures, low dehumidification efficiency, and even compressor mechanical damages.
6. Do not develop/propose unrealizable, doubtful, unworkable (so-called improved) concepts to enforce the credibility of heat pump-assisted drying scientific and industrial communities.
7. Explain how the drying air flow rates are controlled based on both dryer and heat pump physical and thermodynamic parameters, and how outside heat rejection is avoided or minimized.
8. Give more attention to the prevention of hazards (risks) that can occur during drying, this goal could be achieved by providing safe thermodynamic parameters to heat pump compressors to avoid troubleshooting such as short on/off cycles, excessively low or high refrigerant pressures and temperatures, and evaporator frosting.
9. Provide at least the following information: (i) dryer's main characteristics, such as location (hot or cold weather, indoor, outdoor), volume (or mass) capacity, air flow configuration through both the dryer and heat pump (fans, blowers, air vents), and percentage of dryer heat leakage rates (conduction, air exfiltration and infiltration); (ii) for a given heat pump dehumidification

capacity, it is mandatory to determine the required quantity of the dried material to provide high-energy performances); (iii) dryer throughput; (iv) physical properties of wet materials, initial and final moisture contents (dry and/or wet basis), and final quality required; (v) drying schedules for each type of dried material, including the product's initial and required final moisture contents, as well as input and output masses (or volumes), preheating requirements and type of supplementary (backup) heating (if applicable), and for each drying step, air setting dry- and wet-bulb temperatures, or air temperature depression, and air relative humidity; and (vi) heat pump's critical operating parameters (e.g., type of refrigerant used, maximum and minimum allowed discharge or suction pressures and temperatures, drying air flow rate and pattern through both the dryer and heat pump heat exchangers, and heat transfer rates of the heat pump evaporator (dehumidification capacity), condenser (heating capacity), and heat rejection device (when required)).

10. Do not claim that a given heat pump-assisted drying arrangement can successfully dry any kind of product or group of similar products (i.e., agro-food, wood, heat-sensitive, biological, solid, liquid, etc.), in any physical state, within any temperature range and with any drying medium, and without modifying the configuration, operation, and control rules; there is a need for dedicated heat pump-assisted dryers per product or group of similar products (i.e., agro-food, wood, heat-sensitive, biological, solid, liquid, etc.), and for each specific range of drying temperatures.
11. Develop comprehensible control guidelines for each product or group of products that potentially could be dried with heat pump-assisted dryers; applicable range of temperature and moisture content levels have to be clearly identified for each type of dried product, allowing pertinent and reliable future industrial implementations.
12. Specify the durations of preheating (warming-up) steps of materials, drying enclosures (and equipment), and heat sources used.
13. Specify, based on continuous experimental measures, drying air dry- and wet-bulb temperatures, as well as relative and absolute humidity allowing setting running sequences of heat pumps' compressors.
14. Indicate total masses of moisture removed from dried materials.
15. Express modified specific moisture extraction ratios and modified specific energy consumptions based on the system (dryer plus heat pump) energy consumptions during entire drying cycles.
16. Report advantages of heat pumps on drying energy efficiency and final quality of dried products based on rigorous and pertinent theoretical studies or experimental tests, and only after the results were compared with those of conventional convective drying methods for each given product.
17. Validate the claimed advantages of heat pump-assisted drying systems by means of credible modeling/simulation and/or by pertinent laboratory- or field-scale experiments.
18. General statements (or conclusions) such as the followings must be accompanied by indications on how the claimed drying conditions, controls, and performances have been performed: (i) "drying conditions are controlled by adjusting the capacity of the heat pump components"; (ii) "air temperature

in a dryer chamber can be adjusted easily by regulating flow of the refrigerant vapor to the main condenser and heat rejection condenser"; (iii) "drying conditions were controlled by adjusting the capacity of the heat pump components"; and (iv) heat pump technology "allows better control of temperature and relative humidity of the drying process."

19. Do not claim "amazing" (positive) impacts of heat pump-assisted drying systems on final structures, colors, and nutritional qualities of dried products, without experimentally proving how such performances were accomplished and quantified.
20. Explain how "better control of temperature and relative humidity of the drying process" has been achieved when relatively high dehumidifying performances have been reported.
21. Since the complexity of controlling heat pump-assisted systems comes from the requirement to match the material thermal dewatering rate with the surface moisture evaporation rate and the heat pump dehumidification capacity (i.e., heat pump evaporator thermal power) throughout the entire drying cycle, apply one or several methods to continuously or intermittently match the material thermal dewatering rate with the heat pump actual dehumidification rate.
22. Increase the academia-industry multidisciplinary collaboration involving various scientific and engineering pertinent fields.
23. Much effort must be made by academia in close cooperation with industry to develop reliable mathematical models to better understanding heat pump-assisted drying processes.
24. Develop new generation of highly qualified research personnel in heat pump-assisted drying technology.
25. Use artificial intelligence technologies in modeling, simulation, and optimization of drying processes by involving feedback loops to adjust drying parameters based on real-time measurements of product quality.

REFERENCES

Hosseinpour, S., A. Martynenko. 2023. An adaptive fuzzy logic controller for intelligent drying. *Drying Technology* 41:1110–1132.

Lewis, W.K. 1921. The rate of drying of solid materials. *Journal of Industrial and Engineering Chemistry* 13:427–432.

Martynenko, A., A. Mujumdar. 2024. Evolution of control strategies toward intelligent drying. *Drying Technology* 42(4):587–588.

Mujumdar, A.S.; L.X. Huang. 2007 Global R&D needs in drying. *Drying Technology* 25:647–658.

Perré, P. 2024. Artificial intelligence in the field of drying: Revolution or evolution? *Drying Technology* 42(4):589–591.

Sherwood, T.K. 1929. The drying of solids, Part 2. *Industrial and Engineering Chemistry Research Journal* 21:976–980.

Strumillo, C., P. Jones, R. Zylla. 1995. Energy aspects in drying. In: *Handbook of Industrial Drying*, 2nd edition, edited by A.S. Mujumdar, Marcel Dekker, New York, pp. 1241–1275.

Winiczenko, R., K. Gornicki, A. Kaleta, A. Martynenko, M. Janaszek-Mankowska, J. Trajer. 2018. Multi-objective optimization of convective drying of apple cubes. *Computers and Electronics in Agriculture* 145:341–348.

Index

Note: **Bold** and *italic* pages refer to tables and figures, respectively.

G

H

I

J

K

L

M

For Product Safety Concerns and Information please contact our EU representative GPSR@taylorandfrancis.com
Taylor & Francis Verlag GmbH, Kaufingerstraße 24, 80331 München, Germany

www.ingramcontent.com/pod-product-compliance
Ingram Content Group UK Ltd.
Pitfield, Milton Keynes, MK11 3LW, UK
UKHW020202200926
475702UK00004B/581

* 9 7 8 1 0 4 1 0 2 2 5 2 7 *